Lecture Notes in Computer Science 9898

Commenced Publication in 1973
Founding and Former Series Editors:
Gerhard Goos, Juris Hartmanis, and Jan van Leeuwen

More information about this series at http://www.springer.com/series/7409

Magdalena Ortiz · Stefan Schlobach (Eds.)

Web Reasoning and Rule Systems

10th International Conference, RR 2016
Aberdeen, UK, September 9–11, 2016
Proceedings

 Springer

Editors
Magdalena Ortiz
TU Wien
Vienna
Austria

Stefan Schlobach
Computer Science
Vrije Universiteit Amsterdam
Amsterdam, Noord-Holland
The Netherlands

ISSN 0302-9743   ISSN 1611-3349   (electronic)
Lecture Notes in Computer Science
ISBN 978-3-319-45275-3   ISBN 978-3-319-45276-0   (eBook)
DOI 10.1007/978-3-319-45276-0

Library of Congress Control Number: 2016948604

LNCS Sublibrary: SL3 – Information Systems and Applications, incl. Internet/Web, and HCI

Printed on acid-free paper

This Springer imprint is published by Springer Nature
The registered company is Springer International Publishing AG Switzerland

Preface

The growth of the Web is without a doubt one the most far-reaching and transformational changes our world has witnessed in the last decades. It has put at our fingertips amounts of data that were unimaginable until just a couple of decades ago. But owing to the quantity, heterogeneity, and dynamicity of this data, making use of it raises enormous challenges. Managing and accessing Web data calls for increasingly better tools and techniques that are capable of reasoning and can infer useful information from data that may be noisy, distributed, heterogeneous, dynamic, incomplete, and inconsistent. Several successful research efforts have used rule-based systems, which allow us to represent knowledge and draw inferences from it, to overcome these challenges. Extensions and adaptations of classic rule-based languages have found their application in a range of areas like ontologies for the Semantic Web, querying Web data, semantic data management, and common-sense reasoning on the Web.

The International Conference on Web Reasoning and Rule Systems has become a major forum for discussion and dissemination of new results concerning Web reasoning and rule systems. This volume contains the proceedings of the 10th International Conference on Web Reasoning and Rule Systems (RR 2016), held during September 9–11, 2016, in Aberdeen, Scotland. The conference program included keynote talks by Abraham Bernstein, Meghyn Bienvenu, Ian Horrocks, and Leonid Libkin, covering diverse theoretical and practical topics of Web reasoning and rule systems. Extended abstracts of these talks are included in this volume.

The conference program also included presentations of 10 full research papers and three technical communications. The latter are a more concise paper format that provides the opportunity to present preliminary and ongoing work, systems, and applications that are of interest to the RR audience. The accepted papers were selected out of 17 submissions by our Program Committee (PC). This selection was based on four experts reviews (and in one exceptional case, three reviews) for each paper. We are deeply grateful to our PC members for their commitment in the process, and their efforts to provide high-quality constructive feedback to the authors.

To foster the participation and engagement of students, which is fundamental to RR and to our scientific community, RR 2016 hosted a doctoral consortium and a joint poster session, in coordination with the established co-location with the 12th edition of the Reasoning Web Summer School (RW 2016), held just before RR. The generous sponsorship of the NSF was fundamental to these events. The RR Conference and RW Summer School would like to acknowledge the support received from VisitScotland and VisitAberdeenshire, as well as from the Accenture Centre for Innovation and the K-Drive project, for which we are very grateful.

We want to thank the invited speakers for their valuable contribution, and the local organizer Jeff Pan and his team for their hard job organizing this event. We would like to thank our general chair, Umberto Straccia, as well as the doctoral consortium chair, Rafael Peñaloza, our publicity chair, Adila Alfa Krisnadhi, and our sponsorship chair,

Giorgos Stamou. As usual, EasyChair was an excellent conference management system and provided great support for the preparation of these proceedings. Last but not least, we thank all authors and participants of RR 2016, who make this event possible and are the heart of this community; we hope they had a wonderful time in Scotland.

July 2016   Magdalena Ortiz
                                                                              Stefan Schlobach

Organization

General Chair

Umberto Straccia   ISTI-CNR, Italy

Program Chairs

Magdalena Ortiz   TU Wien, Austria
Stefan Schlobach   Vrije Universiteit, Amsterdam, The Netherlands

Doctoral Consortium Chair

Rafael Peñaloza   Free University of Bozen-Bolzano, Italy

Publicity Chair

Adila Krisnadhi   Wright State University, USA and Universitas
     Indonesia, Indonesia

Sponsorship Chair

Giorgos Stamou   NTUA, Greece

Local Chair

Jeff Z. Pan   University of Aberdeen, UK

Local Organising Committee

Wamberto Vasconcelos   University of Aberdeen, UK
Martin Kollingbaum   University of Aberdeen, UK
Diana Zee   University of Aberdeen, UK
Nicola Pearce   University of Aberdeen, UK

Program Committee

Darko Anicic	Siemens AG, Munich, Germany
Meghyn Bienvenu	CNRS, University of Montpellier, Inria, France
Fernando Bobillo	University of Zaragoza, Spain
Elena Botoeva	Free University of Bozen-Bolzano, Italy
Pierre Bourhis	CNRS LIFL/Inria Lille, France
Loris Bozzato	Fondazione Bruno Kessler, Italy
Minh Dao-Tran	TU Wien, Austria
Sergio Flesca	DEIS - University of Calabria, Italy
Paul Fodor	Stony Brook University, USA
Andre Freitas	University of Passau, Germany
Víctor Gutiérrez Basulto	University of Bremen, Germany
André Hernich	University of Liverpool, UK
Aidan Hogan	DCC, Universidad de Chile, Chile
Yazmin Ibanez	TU Wien, Austria
Mark Kaminski	University of Oxford, UK
Benny Kimelfeld	Technion, Israel Institute of Technology, Israel
Roman Kontchakov	Birkbeck, University of London, UK
Markus Krötzsch	Technische Universität Dresden, Germany
Georg Lausen	University of Freiburg, Germany
Joohyung Lee	Arizona State University, USA
Domenico Lembo	Sapienza University of Rome, Italy
Thomas Meyer	Centre for Artificial Intelligence Research, UKZN and CSIR Meraka, South Africa
Marie-Laure Mugnier	University of Montpellier, France
Matthias Nickles	National University of Ireland, Galway, Digital Enterprise Research Institute, Ireland
Andreas Pieris	TU Wien, Austria
Axel Polleres	Vienna University of Economics and Business, Austria
Juan L. Reutter	Pontificia Universidad Católica, Chile
Francesco Ricca	University of Calabria, Italy
Sebastian Rudolph	Technische Universität Dresden, Germany
Vladislav Ryzhikov	Free University of Bozen-Bolzano, Italy
Juan F. Sequeda	Capsenta Labs, Austin, Texas, USA
Evgeny Sherkhonov	University of Amsterdam, The Netherlands
Mantas Simkus	TU Wien, Austria
Daria Stepanova	Max Planck Institute for Informatics, Germany
Domagoj Vrgoc	Pontificia Universidad Católica, Chile
Guohui Xiao	Free University of Bozen-Bolzano, Italy

Additional Reviewers

Alferes, Jose Julio
Güzel, Elem
Hansen, Peter
Schneider, Patrik
Steyskal, Simon
Thomazo, Michaël

EUROPE & SCOTLAND
European Regional Development Fund
Investing in a Smart, Sustainable and Inclusive Future

Contents

Short Papers

On the Complexity of Evaluating Regular Path Queries over Linear Existential Rules

Meghyn Bienvenu[1,2] and Michaël Thomazo[2(✉)]

[1] CNRS, Université de Montpellier, Montpellier, France
meghyn@lirmm.fr
[2] Inria, Le Chesnay Cedex, France
michael.thomazo@inria.fr

Abstract. In the setting of ontology-mediated query answering, a query is evaluated over a knowledge base consisting of a database instance and an ontology. While most work in the area focuses on conjunctive queries, navigational queries are gaining increasing attention. In this paper, we investigate the complexity of evaluating the standard form of navigational queries, namely two-way regular path queries, over knowledge bases whose ontology is expressed by means of linear existential rules. More specifically, we show how to extend an approach developed for DL-Lite$_\mathcal{R}$ to obtain an exponential-time decision procedure for linear rules. We prove that this algorithm achieves optimal worst-case complexity by establishing a matching EXPTIME lower bound.

1 Introduction

Ontology-mediated query answering (OMQA) has generated a lot of interest in the last years as a promising way of facilitating access to data (see [4] for a recent survey). In the OMQA approach, the ontology serves to define a conceptual view of an application domain, introducing a convenient vocabulary for query formulation and providing background knowledge that is exploited at query time to obtain the complete set of answers. So far, the vast majority of research on OMQA has considered user queries in the form of conjunctive queries (CQs), which are a standard query language for relational databases. However, in numerous application scenarios, data can naturally be seen as graphs, in which case so-called *navigational queries* are considered more suitable. The basic navigational query language is regular path queries (RPQs) [11], which allow one to find paths whose labels conform to a given regular language.

In recent years, the problem of answering navigational queries in the setting of OMQA has begun to be explored, first for ontologies formulated in highly expressive description logics (DLs) of the $\mathcal{Z}$ family [8–10], then for rich Horn DLs like Horn-$\mathcal{SROIQ}$ [18], and more recently, for lightweight DLs like DL-Lite$_\mathcal{R}$ and $\mathcal{EL}$ [5,19]. The latter DLs, which underlie the OWL 2 QL and EL profiles, are the most relevant for OMQA due to their favourable computational properties. In addition to plain RPQs, this line of work has also considered richer navigational languages like conjunctive RPQs (which extend both RPQs

M. Ortiz and S. Schlobach (Eds.): RR 2016, LNCS 9898, pp. 1–17, 2016.
DOI: 10.1007/978-3-319-45276-0_1

and CQs) and extensions with nesting and/or negation [3,6,15]. Although much work remains to be done in developing and implementing efficient algorithms, the complexity landscape for answering various forms of path queries over DL knowledge bases is now rather well understood. The same cannot be said for ontologies formulated by means of decidable classes of existential rules (like linear and guarded rulesets), which constitute another important class of ontology languages [1,7]. A key feature that distinguishes existential rules from DLs is the possibility of using predicates of arity greater than two. Since regular path queries are defined only with respect to unary and binary predicates, one might wonder whether they make sense in higher arity settings. We argue however that unary and binary predicates form the backbone of real-world ontologies (irrespective of the choice of ontology language), and it is desirable to be able to use some higher-arity predicates without losing any expressivity in the query language.

In this paper, we take a step towards a better understanding of the combination of navigational query languages and existential rules by studying the complexity of answering two-way RPQs in the presence of linear rules, a well-studied class of existential rules that are a natural generalization of the DL-Lite description logics. After introducing the necessary background, we show how to adapt the RPQ algorithm for DL-Lite proposed in [5] to the setting of linear rules. Unfortunately, our adaptation incurs an exponential blow-up with respect to the maximum predicate arity. We can nevertheless show that the obtained algorithm is worst-case optimal, as RPQ answering is EXPTIME-complete in combined complexity.

2   Preliminaries

We adopt the notation of [13]. The notions of constants, function symbols and predicate symbols are standard. Each function or predicate symbol is associated with a nonnegative integer arity. Variables, terms, substitutions, atoms, first-order formulae, sentences, interpretations (*i.e.*, structures), and models are defined as usual. By a slight abuse of notation, we often identify a conjunction with the set of its conjuncts. Furthermore, we often abbreviate a vector of terms $t_1, \ldots, t_n$ as $\mathbf{t}$, and define $|\mathbf{t}| = n$. By $\varphi\sigma$ we denote the result of applying a substitution σ to φ. A term, atom, or formula is *ground* if it does not contain variables; a *fact* is a ground atom. A term t' is a subterm of a term t if $t' = t$ or $t = f(\mathbf{s})$ where f is a function and t' is a subterm of some $s_i \in \mathbf{s}$. A term s is *contained* in an atom $p(\mathbf{t})$ is $s \in \mathbf{t}$, and s *occurs* in $p(\mathbf{t})$ if s is a subterm of some term $t_i \in \mathbf{t}$; thus, if s is contained in $p(\mathbf{t})$, s occurs in $p(\mathbf{t})$, but the converse may not hold. A term s is *contained* (resp. *occurs*) in a set of atoms I if s is contained (resp. occurs) in some atom in I. Let $T = \{t_1, \ldots, t_n\}$ be a set of terms. A term t is *generated by* T if (i) $t \in T$ or (ii) $t = f(x_1, \ldots, x_k)$ and all the x_k are generated by T. An *instance* is a finite set of function-free facts. The terms appearing in an instance (resp. atom) are denoted by *terms*(I) (resp. *terms*(α)).

Existential Rules. An *existential rule* (or just *rule*) takes the form:

$$\forall \mathbf{x} \forall \mathbf{z}. [\varphi(\mathbf{x}, \mathbf{z}) \rightarrow \exists \mathbf{y}. \psi(\mathbf{x}, \mathbf{y})],$$

where $\varphi(\mathbf{x}, \mathbf{z})$ and $\psi(\mathbf{x}, \mathbf{y})$ are non-empty conjunctions of function-free atoms, and tuples of variables $\mathbf{x}, \mathbf{y}$ and $\mathbf{z}$ are pairwise disjoint. We call φ the *body* and ψ the *head* of the rule. For brevity, quantifiers are often omitted.

We frequently use *Skolemisation* to interpret rules in *Herbrand* interpretations, which are defined as possibly infinite sets of facts. In particular, for each rule ρ and each variable $y_i \in \mathbf{y}$, let f_ρ^i be a function symbol globally unique for ρ and y_i of arity $|\mathbf{x}|$; furthermore, let θ_{sk} be the substitution such that $\theta_{sk}(y_i) = f_\rho^i(\mathbf{x})$ for each $y_i \in \mathbf{y}$. Then, the Skolemisation $sk(\rho)$ of ρ is the following rule: $\varphi(\mathbf{x}, \mathbf{z}) \rightarrow \psi(\mathbf{x}, \mathbf{y})\theta_{sk}$.

A *linear rule* is an existential rule whose body is restricted to a single atom. For ease of presentation, we will consider only rules without any constants. As usual, we also assume that rules have only a single atom in the head. This can be done without loss of generality.

Skolem Chase. The *chase* [14,16] (or canonical model) is a classical tool in OMQA. In this paper, we use the *Skolem chase* variant [17]. Let $\rho = \varphi \rightarrow \psi$ be a Skolemised rule, and let I be a set of facts. A set of facts S is a consequence of ρ on I if a substitution σ exists that maps the variables in ρ to the terms occurring in I (denoted by $terms(I)$) such that $\varphi\sigma \subseteq I$ and $S \subseteq \psi\sigma$. The *result* of applying ρ to I, written $\rho(I)$, is the union of all consequences of ρ on I. If Ω is a set of Skolemised rules, we set $\Omega(I) = \bigcup_{\rho \in \Omega} \rho(I)$. Let I be a finite set of facts, let $\mathcal{R}$ be a set of rules, let $\mathcal{R}' = sk(\mathcal{R})$, and let $\mathcal{R}'_f$ and $\mathcal{R}'_n$ be the subsets of $\mathcal{R}'$ containing rules with and without function symbols, respectively. The *chase sequence* for I and $\mathcal{R}$ is a sequence of sets of facts $I_\mathcal{R}^0, I_\mathcal{R}^1, \ldots$, where $I_\mathcal{R}^0 = I$ and for each $i > 0$, set $I_\mathcal{R}^i$ is defined as follows:

- if $\mathcal{R}'_n(I_\mathcal{R}^{i-1}) \not\subseteq I_\mathcal{R}^{i-1}$, then $I_\mathcal{R}^i = I_\mathcal{R}^{i-1} \cup \mathcal{R}'_n(I_\mathcal{R}^{i-1})$
- otherwise $I_\mathcal{R}^i = I_\mathcal{R}^{i-1} \cup \mathcal{R}'_f(I_\mathcal{R}^{i-1})$

The *chase* of I and $\mathcal{R}$, written $chase(I, \mathcal{R})$, is defined as $\bigcup_i I_\mathcal{R}^i$; note that $chase(I, \mathcal{R})$ can be infinite. However, the chase has a simple structure when linear rules are considered: each atom can be "chased" independently.

Property 1 (Decomposition of the Chase). Let $\mathcal{R}$ be a set of linear rules and I be an instance. It holds that:

$$chase(I, \mathcal{R}) = \bigcup_{\alpha \in I} chase(\{\alpha\}, \mathcal{R})$$

Regular Languages. A regular language can be represented either by a regular expression or by a non-deterministic finite automaton (NFA). Let Σ be a finite set of symbols. A regular expression over Σ is defined by the grammar: $\mathcal{E} \rightarrow \varepsilon \mid a \mid \mathcal{E} \cdot \mathcal{E} \mid \mathcal{E} + \mathcal{E} \mid \mathcal{E}^*$, where $a \in \Sigma$ and ε denotes the empty word. We use $L(\mathcal{E})$ to denote the language defined by $\mathcal{E}$. An NFA over Σ is a tuple $\mathbb{A} = (S, \Sigma, \delta, s_0, F)$, where S is a finite set of states, $\delta \subseteq S \times \Sigma \times S$ is the transition relation, $s_0 \in S$ is the initial state and $F \subseteq S$ is the set of final states. If $\mathbb{A}$ is an automaton and s and s' are two states of $\mathbb{A}$, we denote by $\mathcal{L}_{\mathbb{A}}(s, s')$ the set of words w for which there is path from s to s' in $\mathbb{A}$ labeled by w.

Regular Path Queries. Let $\mathcal{P}$ be a set of predicates. Let us define $\mathcal{P}_2^{\pm} = \mathcal{P}_2 \cup \{r^- \mid r \in \mathcal{P}_2\}$ and $\mathcal{P}_r = \mathcal{P}_2^{\pm} \cup \mathcal{P}_1$, where $\mathcal{P}_i$ ($i \in \{1, 2\}$) denotes the predicates of arity i. A *two-way regular path query* (RPQ[1]) is a query of the form $q(x, x') = \mathcal{E}(x, x')$, where $\mathcal{E}$ is a regular expression defining a language over $\mathcal{P}_r$.

Given an interpretation $\mathcal{I}$, a *path* from a_0 to a_n in $\mathcal{I}$ is a sequence $a_0 r_1 a_1 r_2 \ldots r_n a_n$ such that for any i such that $1 \leq i \leq n$, a_i is an element of the domain $\Delta^{\mathcal{I}}$ of $\mathcal{I}$, every r_i is a symbol from $\mathcal{P}_r$ and:

- if $r_i = a \in \mathcal{P}_1$, then $a_i = a_{i-1} \in a^{\mathcal{I}}$;
- if $r_i \in \mathcal{P}_2$, then $(a_{i-1}, a_i) \in r_i^{\mathcal{I}}$;
- if $r_i = r^-$ with $r \in \mathcal{P}_2$, then $(a_i, a_{i-1}) \in r^{\mathcal{I}}$.

The *label* $\lambda(p)$ of path $p = a_0 r_1 a_1 r_2 \ldots r_n a_n$ is the word $r_1 r_2 \ldots r_n$. For any language L over $\mathcal{P}_r$, the semantics of L with respect to an interpretation $\mathcal{I}$ is defined by:

$$L^{\mathcal{I}} = \{(a_0, a_n) \mid \text{there is some path } p \text{ from } a_0 \text{ to } a_n \text{ such that } \lambda(p) \in L\}.$$

A *match* for an RPQ $q(x, x') = \mathcal{E}(x, x')$ in an interpretation $\mathcal{I}$ is a mapping π from the variables of q to elements of $\Delta^{\mathcal{I}}$ such that $(\pi(x), \pi(x')) \in L(\mathcal{E})^{\mathcal{I}}$.

A *certain answer* to $q(x_1, x_2)$ with respect to $(I, \mathcal{R})$ is a pair of constants (a_1, a_2) such that for every model $\mathcal{I}$ of $(I, \mathcal{R})$, there is a match π for q such that $\pi(x_1) = a_1^{\mathcal{I}}$ and $\pi(x_2) = a_2^{\mathcal{I}}$. As matches are preserved under homomorphisms, it holds that (a_1, a_2) is a certain answer to $q(x_1, x_2)$ w.r.t. $(I, \mathcal{R})$ if and only if there is a match for $(a_1^{\mathcal{I}}, a_2^{\mathcal{I}})$ in $\mathcal{I} = chase(I, \mathcal{R})$. The *RPQ Answering problem* asks, given an RPQ $q(x_1, x_2)$, an instance I, a set of existential rules $\mathcal{R}$, and two constants $(a_1, a_2) \in terms(\mathcal{I}) \times terms(\mathcal{I})$, whether (a_1, a_2) is a certain answer to $q(x_1, x_2)$.

Computational Complexity and Turing Machines. We assume the reader to be familiar with standard complexity classes. In particular, we will consider P, NP, PSpace, APSpace (alternating PSpace), and ExpTime. We recall that APSpace = ExpTime.

To fix notations, we recall that an *alternating Turing machine (TM)* is given by a 5-tuple $\mathcal{M} = (Q, \Gamma, \delta, q_0, g)$ where:

- Q is the finite set of states;
- Γ is the finite tape alphabet;
- $\delta : Q \times \Gamma \to (Q \times \Gamma \times \{L, R\})^2$ is the transition function;
- $q_0 \in Q$ is the initial state;
- $g : Q \to \{\wedge, \vee, \text{accept}, \text{reject}\}$ specifies the type of each state.

---

[1] As we only consider the two-way variant, we will use the abbreviation RPQ instead of the more traditional 2RPQ.

Note that without loss of generality, we consider TMs having the following properties:

- for every universal ($\wedge$) or existential ($\vee$) configuration, there exist exactly two applicable transitions;
- the machine directly accepts any configuration whose state s is such that $g(s) = $ accept;
- the TM never tries to go to the left of the initial position.

We say $\mathcal{M}$ is *polynomially space-bounded* ($\mathcal{M}$ is a PSPACE TM) if there exists a polynomial p such that on input x, $\mathcal{M}$ visits only the first $p(|x|)$ tape cells. We assume w.l.o.g. that the alternating PSPACE TMs we consider terminate on every input.

3   Evaluating Regular Path Queries over Linear Rules

We consider the problem of computing the certain answers to a regular path query and show how to adapt the construction in [5] to the case of linear rules. There are two main ingredients in the original algorithm for DL-Lite:

- a path in the chase is guessed step by step, keeping in memory only the current constant of the instance and current state of the automaton;
- when a path goes through the Skolem part of the chase, these constants are not guessed, but the state in which the automaton is when the path returns to constants of the instance is guessed, thanks to a precomputed table.

3.1   Additional Challenges with Linear Rules

There are two main differences between DL-Lite and linear rules that need to be handled. First, in DL-Lite, it is enough to know the predicate of the atom in which an constant has been created during the chase and the position at which it appeared in that atom to determine all the atoms that contain that constant in the chase. This is not true if we consider general linear rules, as illustrated by the following example:

Example 1 (More Complex Types are Needed). Let us consider the following rules:

$$h(x, y, z) \rightarrow h(z, x, y) \qquad h(x, x, y) \rightarrow q(y)$$

and instance $I = \{h(a, b, b), h(c, d, e)\}$. Observe that while a and c occur in the same position of atoms with the same predicate, $q(a)$ is in $chase(I, \mathcal{R})$, while $q(c)$ is not.

Second, the following looping property is central to the algorithm from [5].

Definition 1 (Looping Property). *An ontology $\mathcal{R}$ fulfills the looping property if it holds that for any instance I, for any path $a_0 r_1 a_1 \ldots r_n a_n$ in $chase(I, \mathcal{R})$ such that* (i) *a_i and a_{i+1} are Skolem terms,* (ii) *a_i is a subterm of a_{i+1}, and* (iii) *a_1 and a_n are original constants, there exists $k \geq i$ such that $a_k = a_i$.*

Indeed, DL-Lite$_\mathcal{R}$ fulfills the looping property (as do many other DLs). However, linear rules do not, as is witnessed by Example 2.

Example 2 (Failure of Looping Property). Consider the instance $I_e = \{t(a, b)\}$ and the ruleset $\mathcal{R}_e$ consisting of the following rules:

$$t(x, y) \rightarrow r(y, z) \qquad\qquad q(x, y, z) \rightarrow p(y, z)$$
$$r(x, y) \rightarrow q(x, y, z) \qquad\qquad q(x, y, z) \rightarrow p(z, x)$$

The chase for I_e and $\mathcal{R}_e$ contains the following atoms:

$$r(b, f_1(b)) \quad q(b, f_1(b), f_2(b, f_1(b))) \quad p(f_1(b), f_2(b, f_1(b))) \quad p(f_2(b, f_1(b)), b)$$

There is thus a path $b \, r \, f_1(b) \, p \, f_2(b, f_1(b)) \, p \, b$ going from the initial constant b to b, that passes by $f_1(b)$ but does not return via $f_1(b)$.

3.2 Adapting the DL-Lite$_\mathcal{R}$ Algorithm

To take care of the first difficulty, we utilize a finer notion of type, which has similar properties to the one used in [5].

Definition 2 (Type). *A type is a pair $(r, \mathcal{P})$ where r is a predicate of arity k and $\mathcal{P}$ is a partition of $\{1, \ldots, k\}$.*

With each atom, we can associate a type, representing the way terms are repeated in the atom.

Definition 3 (Type of an Atom). *Let α be an atom, whose arity is k. The type of α is the pair $(r, \mathcal{P})$ where p is the predicate of α and $\mathcal{P}$ is the partition of $\{1, \ldots, k\}$ such that i and j belong to the same partition iff the ith and the jth arguments of α are equal.*

Note that if two atoms α_1 and α_2 are of same type, there exists an injective substitution θ_{12} such that $\alpha_2 = \alpha_1 \theta_{12}$.

Property 2. Let I be an instance, and $\mathcal{R}$ be a set of linear rules. Let α_1 and α_2 be two atoms of I of same type and θ_{12} such that $\alpha_2 = \alpha_1 \theta_{12}$. Then for every atom β such that $\beta \in chase(\{\alpha_1\}, \mathcal{R})$, $\beta\theta_{12} \in chase(\{\alpha_2\}, \mathcal{R})$.

Let us define for any atom $\alpha \in chase(I, \mathcal{R})$, the restriction of $chase(I, \mathcal{R})$ to α, denoted $chase(I, \mathcal{R})_{|\alpha}$, as the subset of $chase(I, \mathcal{R})$ consisting of those atoms whose terms are generated by $terms(\alpha)$. Observe that by the preceding property, if $type(\alpha) = type(\beta)$, then $chase(I, \mathcal{R})_{|\alpha}$ is isomorphic to $chase(\{\beta\}, \mathcal{R})$.

We can overcome the second difficulty by generalizing the Loop table introduced in [5], which keeps track of the paths that occur 'below' a given type. Intuitively, a type T is in the cell indexed by (s_i, j, s'_i, j') if and only if below any atom of type T, there is a path going from the term in position j to the term in position j' labeled by a word that takes $\mathbb{A}$ from state s_i to state s'_i.

Definition 4 (Loop). *Let $\mathcal{R}$ be a set of linear rules and $\mathbb{A}$ be an NFA. A **Loop** table has cells indexed by tuples $(s_i, j, s_{i'}, j')$ such that s_i and $s_{i'}$ are states of $\mathbb{A}$ and j and j' are integers between 1 and w, where w is the maximum arity appearing in the ruleset. Cells contain types. A **Loop** table is:*

- *sound if for every $T \in (s_i, j, s_{i'}, j')$ it holds that for every atom α of type T appearing in some chase($\{\alpha'\}, \mathcal{R}$) (with the predicate of α' appearing in $\mathcal{R}$), there is a path p in the restriction of chase($I, \mathcal{R}$) to α that goes from argument j of α to argument j' of α such that $\lambda(p) \in \mathcal{L}_{\mathbb{A}}(s_i, s_{i'})$.*
- *complete if for every atom α of type T (whose predicate appears in $\mathcal{R}$), if there is path p from argument j to argument j' of α in chase($\{\alpha\}, \mathcal{R}$) such that $\lambda(p) \in \mathcal{L}_{\mathbb{A}}(s_i, s_{i'})$, then $T \in (s_i, j, s_{i'}, j')$.*

It is direct from the definition that there exists a unique sound and complete Loop table, and in what follows, we use Loop to denote this table.

The table Loop can be constructed using Algorithm 1. Line 5 initializes the table by stating than one can go from a position to the same position without reading any word (and thus not moving in the automaton). Lines 8 and 10 correspond to going through a single edge, reading its label either as an r or an r^-, in the case where both terms are distinct. Lines 13 to 16 do the same thing when both arguments are equal. Line 19 deals with unary predicates. Finally, Lines 23 and 26 saturate the table through respectively transitive closure and propagation of paths from a child to its parent.

Property 3. Let $\mathcal{R}$ be a set of linear rules, I be an instance and $\alpha \in I$. The following are equivalent:

1. $type(\alpha) \in \mathsf{Loop}(s, i, s', j)$
2. there is a path $p = a_0 r_1 a_1 \ldots r_n a_n$ in chase($I, \mathcal{R}$)$_{|\alpha}$ with a_0 appearing at position i in α, a_n appearing at position j in α, and $\lambda(p) \in \mathcal{L}_{\mathbb{A}}(s, s')$.

Proof. ($\Rightarrow$) We prove, by induction on the order of addition of types that whenever a type is added to a cell in $\mathsf{Loop}(s, i, s', j)$, the second condition is fulfilled as well. If $type(\alpha)$ is added to $\mathsf{Loop}(s_i, j, s_i, j)$ at Line 5, the empty word defines a trivial path from any position existing in α to itself, and takes the automaton from any state to itself. If $type(\alpha)$ is added to $\mathsf{Loop}(s_1, 1, s_2, 2)$ at Line 8, α is a binary atom of the form $r(e_1, e_2)$, and there is indeed a path from e_1 to e_2 labeled r. Moreover, there is a transition in $\mathbb{A}$ from s_1 to s_2 labeled by r, which concludes this case. The reasoning is similar for types added via Line 10 and Lines 13 to 16. If $type(\alpha)$ is added at Line 23, it must have already been added to $\mathsf{Loop}(s_1, j_1, s_2, j_2)$ and $\mathsf{Loop}(s_2, j_2, s_3, j_3)$. By the induction assumption, there is a word w_1 (resp. w_2) in $\mathcal{L}_{\mathbb{A}}(s_1, s_2)$ (resp. $\mathcal{L}_{\mathbb{A}}(s_2, s_3)$) that labels a path from the position j_1 (resp. j_2) of an atom α of type T to the position j_2 (resp. j_3). Thus $w_1 \cdot w_2$ labels a path from position j_1 in α to position j_3 in α and belongs to $\mathcal{L}_{\mathbb{A}}(s_1, s_3)$. Finally, let us assume that $type(\alpha)$ is added to $\mathsf{Loop}(s_1, i_{\alpha'}, s_2, j_{\alpha'})$ at Line 26. By assumption, there is a rule $\alpha' \rightarrow \beta'$ in $\mathcal{R}$ such that α and α' have the same type, $type(\beta')$ is in $\mathsf{Loop}(s_1, i_{\beta'}, s_2, j_{\beta'})$, and the same variable

---

Algorithm 1. Creating the Loop table

Data: A set of linear rules $\mathcal{R}$

Result: A sound and complete Loop table

/* Initialization step   */

1 **foreach** *arity* k **do**

2   **foreach** *type* T *of predicate of arity* k **do**

3   **for** $j \in \{1, \ldots, k\}$ **do**

4   **for** $s_i \in Q(\mathbb{A})$ **do**

5                 $\mathrm{Loop}(s_i, j, s_i, j) \leftarrow \mathrm{Loop}(s_i, j, s_j, j) \cup \{T\}$;

6 **for** *type* T *based on* $r(x, y)$ **do**

7   **if** $s_2 \in \delta(s_1, r)$ **then**

8         $\mathrm{Loop}(s_1, 1, s_2, 2) \leftarrow \mathrm{Loop}(s_1, 1, s_2, 2) \cup \{T\}$;

9   **if** $s_2 \in \delta(s_1, r^-)$ **then**

10         $\mathrm{Loop}(s_1, 2, s_2, 1) \leftarrow \mathrm{Loop}(s_1, 2, s_2, 1) \cup \{T\}$;

11 **for** *type* T *based on* $r(x, x)$ **do**

12   **if** $s_2 \in \delta(s_1, r) \cup \delta(s_1, r^-)$ **then**

13         $\mathrm{Loop}(s_1, 1, s_2, 1) \leftarrow \mathrm{Loop}(s_1, 1, s_2, 1) \cup \{T\}$;

14         $\mathrm{Loop}(s_1, 1, s_2, 2) \leftarrow \mathrm{Loop}(s_1, 1, s_2, 2) \cup \{T\}$;

15         $\mathrm{Loop}(s_1, 2, s_2, 1) \leftarrow \mathrm{Loop}(s_1, 2, s_2, 1) \cup \{T\}$;

16         $\mathrm{Loop}(s_1, 2, s_2, 2) \leftarrow \mathrm{Loop}(s_1, 2, s_2, 2) \cup \{T\}$;

17 **for** *type* T *based on* $a(x)$ **do**

18   **if** $s_2 \in \delta(s_1, a)$ **then**

19         $\mathrm{Loop}(s_1, 1, s_2, 1) \leftarrow \mathrm{Loop}(s_1, 1, s_2, 1) \cup \{T\}$;

/* Saturation step   */

20 **while** *something added* **do**

21   **for** T *a type* **do**

22   **if** $T \in \mathit{Loop}(s_1, j_1, s_2, j_2) \cap \mathit{Loop}(s_2, j_2, s_3, j_3)$ **then**

23             $\mathrm{Loop}(s_1, j_1, s_3, j_3) \leftarrow \mathrm{Loop}(s_1, j_1, s_3, j_3) \cup \{T\}$;

24   **for** $\alpha \rightarrow \beta \in \mathcal{R}$, *of respective types* T_α, T_β **do**

25   **if** *the same variable appears in* α *at* i_α *and* β *at* i_β *(resp.* j_α *and* j_β),
        $T_\beta \in \mathit{Loop}(s_1, i_\beta, s_2, j_\beta)$ **then**

26             $\mathrm{Loop}(s_1, i_\alpha, s_2, j_\alpha) \leftarrow \mathrm{Loop}(s_1, i_\alpha, s_2, j_\alpha) \cup \{T_\alpha\}$;

---

appears at position $i_{\alpha'}$ (resp. $j_{\alpha'}$) in α' and $i_{\beta'}$ (res. $j_{\beta'}$) in β'. By the induction assumption, there is a word $w \in \mathcal{L}_\mathbb{A}(s_1, s_2)$ that labels a path from $i_{\beta'}$ to $j_{\beta'}$. Now, let us observe that any two terms that are at positions $i_{\alpha'}$ and $j_{\alpha'}$ of the same atom of type $type(\alpha')$ are also at position $i_{\beta'}$ and $j_{\beta'}$ of an atom of type $type(\beta')$ in $chase(D, \mathcal{R})_{|\alpha}$ because it is a model of $\alpha' \rightarrow \beta'$. Thus, w is also the label of a path from the term at position i'_α to the term at position j'_α, which concludes the proof.

($\Leftarrow$) We suppose that the second statement holds and reason by induction on the length n of the path $p = a_0 r_1 a_1 \ldots r_n a_n$.

Base case, path of length 0: both states and database constants are thus equal, and the type is added by the initialization in Line 5.

Base case, path of length 1: $\alpha' = r_1(a_0, a_1)$ belongs to $chase(I, \mathcal{R})_{|\alpha}$, and $r_1 \in \mathcal{L}_{\mathbb{A}}(s, s')$. If $a_0 \neq a_1$, then $type(\alpha')$ is added to the cells $(s, 1, s', 2)$ and $(s, 1, s', 2)$ in Lines 8 and 10. If $a_0 = a_1$, then $type(\alpha')$ is added to the four cells (s, i', s', j') with $i', j' \in \{1, 2\}$ (Lines 13–16). As α' belongs to $chase(I, \mathcal{R})_{|\alpha}$, there exists a finite sequence of atoms $\alpha = \alpha_0, \ldots, \alpha_m = \alpha'$ such that α_{i+1} belongs to $\rho_i(\alpha_i)$ for some rule $\rho_i \in \mathcal{R}$. By using m applications of Line 26, we obtain $type(\alpha) \in \mathsf{Loop}(s, i, s', j)$.

Induction step: let us assume that the result holds for any path of length up to $n - 1, n \geq 2$, and consider the path $p = a_0 r_1 a_1 \ldots r_n a_n$. First consider the case in which a_k is contained in α for some $1 \leq k < n$, and let l be a position of a_k in α. There exists a path from a_0 to a_k of length strictly smaller than n, and similarly from a_k to a_n. By the induction assumption, $type(\alpha)$ is in both $\mathsf{Loop}(s, i, s'', l)$ and $\mathsf{Loop}(s'', l, s', j)$ for some state s''. An application of Line 23 yields $type(\alpha) \in \mathsf{Loop}(s, i, s', j)$. Next suppose there is no a_k ($1 \leq k < n$) that occurs in α, and let β be the atom in which a_1 is created (at position k'). This atom is well defined as we consider rules with atomic head. We know that a_0 (resp. a_n) must occur in β, let us say at position i' (resp. j'). Indeed, if it was not the case, α should contain a term among $a_1, \ldots, a_{n-1}$ which contradicts our earlier assumption. By the induction hypothesis, $type(\beta)$ belongs to $\mathsf{Loop}(s, i', s'', k')$ and to $\mathsf{Loop}(s'', k', s', j')$ for some state s''. Hence, by Line 23, $type(\beta)$ is in the cell $\mathsf{Loop}(s, i', s', j')$. By (repeated) application of Line 26, $type(\alpha)$ is in the cell $\mathsf{Loop}(s, i, s', j)$, which concludes the proof.   □

Property 4. Algorithm 1 runs in exponential time, and in polynomial time if the predicate arity is bounded.

Proof. There are polynomially many cells in the table, each of which can contain at most all types. The number n_t of distinct types is single exponential (and polynomial for bounded-arity predicates). The first for loop runs in $\mathcal{O}(n_t)$, the next two run in polynomial time, and the while loop is performed at most n_t times.   □

The remainder of the decision procedure is very close to the original algorithm for DL-Lite$_{\mathcal{R}}$, but we recall it here (Algorithm 2) in the interest of self-containment. The idea is as follows: starting from a constant a and the initial state of $\mathbb{A}$, we guess the next constant in I on a path from a to b and the state of $\mathbb{A}$ after taking this step (Line 7). We then check that this choice is valid, i.e., there is indeed a path from a to the guessed constant which takes the automaton from the initial state to the current guessed state. This can be done either by a checking that a corresponding unary or binary atom is entailed (Lines 9 and 10), or by checking that a path going through the Skolem part of the chase allows us to reach the next constant in the required state, using the Loop table (Lines 12 to 14). We repeat this procedure until we reach the constant b in a final state, or hit the maximal path length. Note that at Line 12, α is uniquely defined if

---

Algorithm 2. RPQ answering over linear rules

Input: An NFA $\mathbb{A}$, an instance I, a set of linear rules $\mathcal{R}$,
  $(a, b) \in terms(I) \times terms(I)$
Output: Yes if and only if (a, b) is a certain answer to the query q defined by $\mathbb{A}$

1 **if** $(I, \mathcal{R})$ *is not satisfiable* **then**
2 | **return** *Yes*

3 current $= (a, s_0)$;
4 count $= 0$, max $= |\mathbb{A}| \times |I|$;
5 **while** *count* $<$ *max and* current $\notin \{(b, s_f) \mid s_f \in F\}$ **do**
6 | Define $(c, s) = $ current;
7 | Guess (d, s') together with $(s, \sigma, s') \in \delta$ or T, i_c, i_d such that
   | $T \in \mathsf{Loop}(s, i_c, s', i_d)$;
8 | **if** (s, σ, s') *was guessed* **then**
9 | | **if** $\sigma \in \mathcal{P}_2^{\pm} \wedge (I, \mathcal{R} \not\models \sigma(c, d))$ **then return** *No*;
10 | | **if** $\sigma = A \wedge (c \neq d \vee I, \mathcal{R} \not\models A(c))$ **then return** *No*;
11 | **if** T, i_c, i_d *was guessed* **then**
12 | | Let α be of type T such that c is at position i_c and d is at position i_d;
   | | other terms are set to fresh variables
13 | | **if** α *does not exist* **then return** *No*;
14 | | **if** $I, \mathcal{R} \not\models \alpha$ **then return** *No*;
15 | current $= (d, s')$, count $=$ count $+1$;

16 **if** *current* $= (b, s_f)$ *for some* $s_f \in F$ **then return** *Yes* **else return** *No*;

---

it exists (it may not exist e.g., if c and d are different but are at positions that should have identical terms according to T).

The following property will be used to establish correctness of the algorithm.

Property 5. At the beginning of each iteration of the while loop of Algorithm 2, it holds that there is a path from a to the first element of current that takes the NFA $\mathbb{A}$ from the initial state s_0 to the state in the second argument of current.

Proof. At the beginning of the first iteration of the while loop, current is equal to (a, s_0). Thus, the path a, whose label is ε, goes from a to a and $\varepsilon \in \mathcal{L}_{\mathbb{A}}(s_0, s_0)$.

Let (a_i, s_i) be the content of current at the beginning of the ith iteration of the while loop. Let w_i be the label of a path from a_0 to a_i such that $w_i \in \mathcal{L}_{\mathbb{A}}(s_0, s_i)$. If there is an $(i + 1)^{\text{th}}$ iteration, either (s, σ, s') or (T, i_c, i_d) has been guessed, and the corresponding check was successful. Let us consider each case:

– if (s, σ, s') has been guessed and checked, we have two cases:
  • $\sigma \in \mathcal{P}_2^{\pm}$, and there is a path from a_i to a_{i+1} in $chase(I, \mathcal{R})$ labeled by σ. Moreover, σ labels an edge from s to s' in $\mathbb{A}$. We can thus define $w_{i+1} = w_i.\sigma$
  • $\sigma = A$, and $I, \mathcal{R} \models A(c)$. As $c = d$, we can again define $w_{i+1} = w_i.\sigma$
– if (T, i_c, i_d) has been guessed, it means that T belongs to $\mathsf{Loop}(s_i, i_c, s_{i+1}, i_d)$. By the definition of Loop, there is a path p (in the Skolem part) from any term at position i_c of an atom of type T to the position i_d of an atom of type

T such that $\lambda(p) \in \mathcal{L}_{\mathbb{A}}(s, s')$. Let α be as defined Line 12. As $I, \mathcal{R} \models \alpha$, where $type(\alpha) = T$, a_i appears at position i_c of α, and a_{i+1} appears at position i_d of α, there is such a path from a_i to a_{i+1}. We can thus set $w_{i+1} = w_i.p$.   □

Property 6. There is an execution of Algorithm 2 that outputs Yes iff the RPQ given by $\mathbb{A}$ is entailed from $(I, \mathcal{R})$.

Proof. ($\Rightarrow$) If the algorithm outputs Yes, the while loop has been exited with `current` equal to (b, s_f), with s_f a final state of $\mathbb{A}$. By Property 5, this means that there is a path from a to b whose label takes $\mathbb{A}$ from s_0 to s_f, hence is accepted by $\mathbb{A}$. This show that whenever Algorithm 2 accepts, (a, b) is a certain answer to the RPQ given by $\mathbb{A}$.

($\Leftarrow$) If (a, b) is a certain answer to the RPQ based upon $\mathbb{A}$, then there is path of minimal length $p = a'_0 r_1 a'_1 \ldots r_n a'_n$ from $a = a'_0$ to $b = a'_n$ in $chase(I, \mathcal{R})$ such that $\lambda(p) = r_1 \ldots r_n \in \mathcal{L}_{\mathbb{A}}(s_0, s_f)$ for some final state s_f. Let $s'_0 s'_1 \ldots s'_n$ be a sequence of states of $\mathbb{A}$ such that s'_n is a final state of $\mathbb{A}$ and for every $1 \leq i \leq n$, $(s_{i-1}, r_i, s_i) \in \delta$. Since p is of minimal length, there is no pair (i, j) with $i \neq j$ such that $(a_i, s_i) = (a_j, s_j)$. Let us consider the sequence $p' = ((a_i, s_i))_i$ such that:

- for any i, a_i is the ith constant, say a'_{k_i}, in p belonging to $terms(I)$;
- for any i, $s_i = s'_{k_i}$.

Moreover, for any i, if $k_{i+1} = k_i + 1$, we define $aux_i = (s_i, r_{i+1}, s_{i+1})$. Otherwise, let $aux_i = (type(\alpha), i_c, i_d)$, where:

- α is such that $\alpha \in I$ and $type(\alpha) \in \text{Loop}(s_i, i_c, s_{i+1}, i_d)$;
- a_{k_i} appears at position i_c of α and $a_{k_{i+1}}$ appears at position i_d of α.

In the second case, it is possible to define aux_i in such a way, as the path $p_s = a'_{k_i} r_{k_i+1} \ldots a'_{k_{i+1}}$ goes from a_{k_i} to $a_{k_{i+1}}$ and belongs to $\mathcal{L}_{\mathbb{A}}(s_i, s_{i+1})$ by definition of s_i. We show that the sequence of guesses (a_i, s_i, aux_i) leads Algorithm 2 to accept. Since p is minimal, the length of p' is less than $|\mathbb{A}| \times |I|$. Moreover, $a_n = b$ and s_f is a final state. Thus, the only way for Algorithm 2 to reject with this sequence of guesses is to reject during checks, *i.e.*, one of the checks performed at Lines 9, 10, 12 or 14 fails. Let (a_i, s_i, aux_i) be the guess at one of the steps. If aux_i is of the form (s_i, r_{i+1}, s_{i+1}), then a_{k_i} and $a_{k_{i+1}}$ are consecutive elements in p, and there is an atom $r_{i+1}(a_{k_i}, a_{k_{i+1}})$ in $chase(I, \mathcal{R})$. Thus, $r_{i+1}(a_{k_i}, a_{k_{i+1}})$ is entailed by I and $\mathcal{R}$, and the check at Line 9 or 10 (depending on r_{i+1} being a binary or unary atom) is successful. If aux_i is of the form $(type(\alpha), i_c, i_d)$, then there is $\alpha \in I$ such that $type(\alpha) \in \text{Loop}(s_i, i_c, s_{i+1}, i_d)$, and with a_{k_i} (resp. $a_{k_{i+1}}$) appearing at position i_c (resp. i_d) of α. The atom α fulfills the conditions of Lines 12 and 14. Thus the defined sequence never triggers a rejection from Algorithm 2, which concludes the proof.   □

Theorem 1. RPQ *Answering in the presence of linear existential rules is:*

- *in* NL *in data complexity*
- *in* PTime *in combined complexity with bounded arity*
- *in* ExpTime *in combined complexity with unbounded arity*

Proof. Algorithm 2 is a non-deterministic algorithm that needs to keep in memory the current state, the current constant, and the number of iterations done so far. It performs two types of operations: entailment checks and accessing the contents of the Loop table (more precisely, deciding whether $T \in \mathsf{Loop}(s, i_c, s', i_d)$). Hence, it can be seen as an NL algorithm making oracle calls whenever an entailment check is performed or a cell of Loop is retrieved. Entailment checks are in NL in data complexity, and Loop is independent from the data: the overall algorithm thus runs in NL in data complexity. In combined complexity with bounded arity, entailment checks can be performed in PTime, while Loop can be computed in polynomial time: the overall algorithm is thus in PTime with bounded arity. In the unbounded arity case, the entailment checks can be performed in PSpace, while the Loop table can be computed in ExpTime: the algorithm thus runs in ExpTime. □

4   Lower Bound

It is already known that the data complexity (resp. combined complexity) of RPQs under linear rules (resp. linear rules with bounded arity) is NL-hard (resp. PTime-hard) [5], which matches the upper bounds obtained in the preceding section. We thus focus on providing a matching ExpTime lower bound for the combined complexity of evaluating RPQs under linear rules of unbounded arity. The proof is done by simulating an alternating PSpace TM. It is already known that PSpace TMs can be simulated by means of linear rules [12]. In the following, we explain how to adapt this construction to simulate alternating TMs. Note that in this section, we will use rules with multiple atoms in the head: this is done to simplify the presentation, and a classical transformation allows us to get the same lower bound for rules with atomic heads.

The intuition is as follows: the construction in [12] represents the configuration of a TM $\mathcal{M}$ by a single atom of polynomial arity. The initial configuration can thus be represented by an instance $I_{\mathcal{M}}$ containing a single atom. Then, for each transition of the TM, polynomially many linear rules are created, each one representing the action of the transition on a cell at a given position. All these rules are part of $\mathcal{R}_{\mathcal{M}}$. The initial configuration of the TM is accepted if and only if an atom encoding a configuration having an accepting state is entailed by $I_{\mathcal{M}}$ and $\mathcal{R}_{\mathcal{M}}$.

We modify this construction in the following way to deal with alternating Turing machines: to each atom, we add two positions, that will act as "input" and "output" positions. Moreover, we will maintain the following property: there is a path, whose edges are all labeled by the same predicate p, from the input position of α to the output position of α entailed by $chase(I_{\{\alpha\}}, \mathcal{R}_{\mathcal{M}})$ if and only if the configuration represented by α is accepted by $\mathcal{M}$. This is true in the following cases:

- the state of the current configuration is accepting. It is then enough to add a p-edge from i_c to o_c; this is possible as the Turing machine is assumed to never leave an accepting state;
- the current state is existential and one of the two successor configurations is accepting: we thus add p-edges from the input of the current configuration to the input of the two children, and from the output of the two children to the output of the current configuration;
- the current state is universal, and both successor configurations are accepting: we thus add p-edges from the input of the current configuration to the input of the first successor configuration, then from the output of that configuration to the input of the other successor, and lastly from the output of the second successor to the output of the current configuration.

We now formalize the construction sketched above, staying as close as possible to the notations in [12].

Turing Machine. Given an alternating PSPACE TM and an input x, we can represent a configuration c reached during the computation by storing the content of the first $p(|x|)$ cells, as well as the position of the head of the tape and the current state of the TM. Adding input and output positions, this can be encoded by a predicate *conf* of arity $2p(|x|) + 3$:

$$\text{conf}(i_c, \textit{state}, \textit{cell}_1, \textit{cur}_1, \textit{cell}_2, \textit{cur}_2, \ldots, \textit{cell}_{p(|x|)}, \textit{cur}_{p(|x|)}, o_c),$$

where *state* contains the state identifier, $\textit{cell}_i$ represents the content of the ith cell, $\textit{cur}_i$ is equal to 1 if the head of the Turing machine is on cell i and 0 otherwise, and i_c and o_c are the input and output terms of this atom. We say that the above atom *represents* configuration c. Given an atom α, the term at its input (resp. output) position is denoted by $i(\alpha)$ (resp. $o(\alpha)$). We denote by $I_{\mathcal{M},x}$ the instance containing a single atom representing the initial configuration of $\mathcal{M}$ on input x.

For every state q_f with $g(q_f) = \text{accept}$, we create the following rule:

$$\text{conf}(i_c, q_f, \ldots, o_c) \rightarrow p(i_c, o_c). \tag{1}$$

For each transition $\delta(q, \gamma) = \{(q', \gamma', L), (q'', \gamma'', L)\}$ such that $g(q) = \vee$, we create the rule

$$\text{conf}(i_c, q, \textit{cell}_1, \textit{cur}_1, \ldots, \textit{cell}_{i-1}, 0, \gamma, 1, \ldots, o_c) \rightarrow$$
$$\exists i_{c'}, o_{c'}, i_{c''}, o_{c''} \; \text{conf}(i_{c'}, q', \textit{cell}_1, \textit{cur}_1, \ldots, \textit{cell}_{i-1}, 1, \gamma', 0, \ldots, o_{c'}),$$
$$\text{conf}(i_{c''}, q'', \textit{cell}_1, \textit{cur}_1, \ldots, \textit{cell}_{i-1}, 1, \gamma', 0, \ldots, o_{c''}),$$
$$p(i_c, i_{c'}), p(o_{c'}, o_c), p(i_c, i_{c''}), p(o_{c''}, o_c). \tag{2}$$

for each position i on the tape, and similarly when the head is moving to the right.

When $g(q) = \wedge$, we associate with each transition $\delta(q, \gamma) = \{(q', \gamma', L),$ $(q'', \gamma'', L)\}$ the following rule:

$$conf(i_c, q, cell_1, cur_1, \ldots, cell_i, 0, \gamma, 1, \ldots, o_c) \rightarrow$$
$$\exists i_{c'}, o_{c'}, i_{c''}, o_{c''} \; conf(i_{c'}, q', cell_1, cur_1, \ldots, cell_i, 1, \gamma', 0, \ldots, o_{c'}),$$
$$conf(i_{c''}, q'', cell_1, cur_1, \ldots, cell_i, 1, \gamma'', 0, \ldots, o_{c''}),$$
$$p(i_c, i_{c'}), p(o_{c'}, i_{c''}), p(o_{c''}, o_c). \quad (3)$$

Figure 1 illustrates the functioning of rules of types (2) and (3). We denote by $\mathcal{R}_{\mathcal{M},x}$ the set containing all the rules defined above[2]. The above rules (where input and output positions are removed) simulate the run of a PSPACE TM [12].

The following property formalizes the reduction and establishes its correctness.

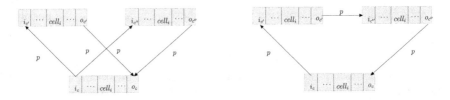

Fig. 1. Existential (left) and universal (right) gadgets

Property 7. Let $\mathcal{M}$ be an alternating PSPACE Turing machine, and let α be an atom of $chase(I_{\mathcal{M},x}, \mathcal{R}_{\mathcal{M},x})$ representing a configuration $c(\alpha)$. Then $c(\alpha)$ is an accepting configuration of $\mathcal{M}$ if and only if there is a path in $chase(I_{\mathcal{M},x}, \mathcal{R}_{\mathcal{M},x})$ from $i(\alpha)$ to $o(\alpha)$ whose label belongs to p^*.

Proof. ($\Leftarrow$) Let $\alpha \in chase(I_{\mathcal{M},x}, \mathcal{R}_{\mathcal{M},x})$ represent a configuration $c(\alpha)$, and let C_α be the restriction of $chase(I_{\mathcal{M},x}, \mathcal{R}_{\mathcal{M},x})$ to α. We show by induction on the number of atoms of C_α that the required path exists. Note that the induction is well-founded as the Skolem chase is finite (recall that the considered Turing machines terminate).

- If C_α contains one atom, then there can be no path in $chase(I_{\mathcal{M},x}, \mathcal{R}_{\mathcal{M},x})$ witnessing $p^*(i(\alpha), o(\alpha))$. Suppose then that C_α contains two atoms. In this case, the only atom in C_α other than α must be $p(i(\alpha), o(\alpha))$. The only way to derive such an atom is to apply a rule of the form (1), which is applied if and only if $c(\alpha)$ is in an accepting state, hence $c(\alpha)$ is an accepting configuration of $\mathcal{M}$.
- Next assume that the result holds for any atom α such that C_α has less than n atoms, and let α be an atom such that C_α contains n atoms. We distinguish two cases:

---

[2] Note that x is required to determine the arity of *conf*.

- Case 1: the state of $c(\alpha)$ is existential. Then, since the rules of type (2) must be satisfied, C_α contains atoms α_1 and α_2 representing the successor configurations of $c(\alpha)$. The existence of a path from $i(\alpha)$ to $o(\alpha)$ implies that there is either a path from $i(\alpha_1)$ to $o(\alpha_1)$ or a path from $i(\alpha_2)$ to $o(\alpha_2)$. To see why, observe that every p-atom involving $i(\alpha)$ or $o(\alpha)$ is added either by the same rule application as created α or by a rule of type (2) applied to α. Only atoms of the second kind (refer to Fig. 1, left) can belong to a shortest path from $i(\alpha)$ to $o(\alpha)$, as atoms of the first kind have $i(\alpha)$ (resp. $o(\alpha)$) as second (resp. first) argument. If we have a path from $i(\alpha_1)$ to $o(\alpha_1)$, then we can apply the induction assumption to α_1 to get that $c(\alpha_1)$ is an accepting configuration, which implies that $c(\alpha)$ is also accepting. We can proceed analogously if we have path from $i(\alpha_2)$ to $o(\alpha_2)$.
- Case 2: the state of $c(\alpha)$ is universal. As the rules of type (3) must be satisfied, the existence of a path from $i(\alpha)$ to $o(\alpha)$ implies the existence of a path from $i(\alpha_1)$ to $o(\alpha_1)$ and a path from $i(\alpha_2)$ to $o(\alpha_2)$, where α_1 and α_2 represent the successor configurations of $c(\alpha)$ (refer to Fig. 1, right). By the induction assumption, $c(\alpha_1)$ and $c(\alpha_2)$ are both accepting configurations, which means that $c(\alpha)$ is also accepting.

($\Rightarrow$) We prove the other direction by induction on the number of transitions that need to be performed to prove that $c(\alpha)$ is accepted by $\mathcal{M}$.

- If no transitions are required, this means that $c(\alpha)$ is in an accepting state. Thus, Rule (1) is applicable, and $p(i(\alpha), o(\alpha))$ is present in $chase(I_{\mathcal{M},x}, \mathcal{R}_{\mathcal{M},x})$.
- Assume the result holds up to n required transitions. We distinguish two cases:
  - Case 1: the state of $c(\alpha)$ is existential. As $c(\alpha)$ is accepting, this means that one of its two successor configurations, say $c(\alpha_1)$, is accepting. Moreover, the number of transitions required to accept $c(\alpha_1)$ is strictly smaller than for $c(\alpha)$. By the induction assumption, $p^*(i(\alpha_1), o(\alpha_1))$ is present in $chase(I_{\mathcal{M},x}, \mathcal{R}_{\mathcal{M},x})$. As $p(i(\alpha), i(\alpha_1))$ and $p(o(\alpha_1), o(\alpha))$ are also present (since the rules of the form (2) generate them), this proves that $p^*(i(\alpha), o(\alpha))$ is present as well.
  - Case 2: the state of $c(\alpha)$ is universal. As $c(\alpha)$ is accepting, this means that its two successor configuration are also accepting. By the induction assumption, this means that $p^*(i(\alpha_1), o(\alpha_1))$ and $p^*(i(\alpha_2), o(\alpha_2))$ are present in $chase(I_{\mathcal{M},x}, \mathcal{R}_{\mathcal{M},x})$. As the rules of the form (3) also generate $p(i(\alpha), i(\alpha_1))$, $p(o(\alpha_1), i(\alpha_2))$, and $p(o(\alpha_2), o(\alpha))$, this proves that $p^*(i(\alpha), o(\alpha))$ is present in $chase(I_{\mathcal{M},x}, \mathcal{R}_{\mathcal{M},x})$. □

Now let $\mathcal{M}$ be an alternating PSPACE Turing machine, x be an input to $\mathcal{M}$, and α be the unique atom in $I_{\mathcal{M},x}$. Then by Property 7, $c(\alpha)$ is an accepting configuration of $\mathcal{M}$ if and only if $I_{\mathcal{M},x}, \mathcal{R}_{\mathcal{M},x} \models p^*(i(\alpha), o(\alpha))$. This, together with known results, yields the following lower bounds:

Theorem 2. RPQ *Answering in the presence of linear existential rules is* NL-*hard in data complexity,* PTIME-*hard in combined complexity with bounded arity and* EXPTIME-*hard in combined complexity without arity bound, even for a fixed RPQ.*

Note that the preceding reduction can be easily adapted to show that atomic query answering under rulesets containing linear rules and transitivity rules is EXPTIME-hard. Assuming EXPTIME≠PSPACE, this result is in contradiction with Theorem 5 in [2], which purports to show a PSPACE upper bound. Indeed, after reexamining the proofs, the authors of the latter work have identified the flaw, which occurs in the analysis of the combined complexity of their rewriting-based decision procedure. It turns out that the procedure runs in exponential time, rather than in polynomial space (the NL upper bound in data complexity remains valid). Combining our lower bound with their procedure shows that the problem is EXPTIME-complete in combined complexity.

5 Conclusion and Future Work

In this paper, we have investigated the complexity of evaluating regular path queries under linear existential rules. We have shown that it is NL-complete in data complexity, PTIME-complete in combined complexity when the predicate arity is bounded, and EXPTIME-complete otherwise. This behavior is somewhat surprising with respect to prior work: indeed, for DL-Lite$_\mathcal{R}$, the combined complexity of RPQ answering is lower than for CQs, whereas we observe just the opposite in the linear case (recall CQ answering is PSPACE-complete under linear rules). The upper bound was shown by adapting an existing decision procedure for DL-Lite, using a refined definition of type. The lower bound builds upon a PSPACE-hardness result for CQ answering under linear rules.

There are two natural ways to extend the present work: either investigate more expressive forms of path queries (with conjunction and/or nesting) over linear rules, or consider the effect of moving to more expressive decidable classes of existential rules.

Acknowledgements. This work was supported by the ANR project 12 JS02 007 01.

References

1. Baget, J., Leclère, M., Mugnier, M., Salvat, E.: Extending decidable cases for rules with existential variables. In: Proceedings of IJCAI, pp. 677–682 (2009)
2. Baget, J., Bienvenu, M., Mugnier, M., Rocher, S.: Combining existential rules and transitivity: next steps. In: Proceedings of IJCAI, pp. 2720–2726 (2015)
3. Bienvenu, M., Calvanese, D., Ortiz, M., Šimkus, M.: Nested regular path queries in description logics. In: Proceedings of KR (2014)
4. Bienvenu, M., Ortiz, M.: Ontology-mediated query answering with data-tractable description logics. In: Faber, W., Paschke, A. (eds.) Reasoning Web 2015. LNCS, vol. 9203, pp. 218–307. Springer, Heidelberg (2015)

5. Bienvenu, M., Ortiz, M., Simkus, M.: Regular path queries in lightweight description logics: complexity and algorithms. J. Artif. Intell. Res. (JAIR) **53**, 315–374 (2015)
6. Bourhis, P., Krötzsch, M., Rudolph, S.: How to best nest regular path queries. In: Proceedings of DL, pp. 404–415 (2014)
7. Calì, A., Gottlob, G., Kifer, M.: Taming the infinite chase: query answering under expressive relational constraints. In: Proceedings of KR, pp. 70–80 (2008)
8. Calvanese, D., Eiter, T., Ortiz, M.: Answering regular path queries in expressive description logics: an automata-theoretic approach. In: Proceedings of AAAI, pp. 391–396 (2007)
9. Calvanese, D., Eiter, T., Ortiz, M.: Regular path queries in expressive description logics with nominals. In: Proceedings of IJCAI, pp. 714–720 (2009)
10. Calvanese, D., Eiter, T., Ortiz, M.: Answering regular path queries in expressive description logics via alternating tree-automata. Inf. Comput. **237**, 12–55 (2014)
11. Florescu, D., Levy, A., Suciu, D.: Query containment for conjunctive queries with regular expressions. In: Proceedings of PODS (1998)
12. Gottlob, G., Papadimitriou, C.H.: On the complexity of single-rule datalog queries. Inf. Comput. **183**(1), 104–122 (2003)
13. Grau, B.C., Horrocks, I., Krötzsch, M., Kupke, C., Magka, D., Motik, B., Wang, Z.: Acyclicity notions for existential rules and their application to query answering in ontologies. J. Artif. Intell. Res. (JAIR) **47**, 741–808 (2013)
14. Johnson, D.S., Klug, A.C.: Testing containment of conjunctive queries under functional and inclusion dependencies. J. Comput. Syst. Sci. **28**(1), 167–189 (1984)
15. Kostylev, E.V., Reutter, J.L., Vrgoc, D.: XPath for DL ontologies. In: Proceedings of AAAI (2015)
16. Maier, D., Mendelzon, A.O., Sagiv, Y.: Testing implications of data dependencies. ACM Trans. Database Syst. **4**(4), 455–469 (1979)
17. Marnette, B.: Generalized schema-mappings: from termination to tractability. In: Proceedings of PODS, pp. 13–22 (2009)
18. Ortiz, M., Rudolph, S., Šimkus, M.: Query answering in the Horn fragments of the description logics $\mathcal{SHOIQ}$ and $\mathcal{SROIQ}$. In: Proceedings of IJCAI (2011)
19. Stefanoni, G., Motik, B., Krötzsch, M., Rudolph, S.: The complexity of answering conjunctive and navigational queries over OWL 2 EL knowledge bases. J. Artif. Intell. Res. (JAIR) **51**, 645–705 (2014)

Towards Practical OBDA
with Temporal Ontologies
(Position Paper)

Diego Calvanese, Elem Güzel Kalaycı[(✉)], Vladislav Ryzhikov,
and Guohui Xiao

KRDB Research Centre for Knowledge and Data, Free University
of Bozen-Bolzano, Bolzano, Italy
{calvanese,kalayci,ryzhikov,xiao}@inf.unibz.it

Abstract. The temporal dimension of data, which contains such important information as duration or sequence of events and is present in many applications of ontology-based data access (OBDA) concerned with logs or streams, is getting growing attention in the community. To give a proper treatment to the events occurring in the data from the ontological perspective, we assume in our approach that every concept is temporalized, i.e., has temporal validity time, and the ontology language expresses the constraints between validity times of concepts. In this paper we outline the state of art and the future challenges of our research. On the theoretical side, we are interested in enriching the ontology languages with the operators for constructing the temporal concepts that are expressive enough to capture the patterns required by industrial use-cases. On the practical side, we are interested in implementing the ontology-mediated query answering with temporalized concepts in the OBDA system Ontop and performing extensive evaluations using large amounts of real-world data.

Keywords: Ontology-based data access · Temporal logic · Description logic

1 Introduction and Motivation

Ontology-based data access (OBDA) [9,21], one of the most promising applications of Knowledge Representation in the Semantic Web area, exposes a high level conceptual layer in the form of an ontology on top of (potentially very large and heterogeneous) data sources. The conceptual view of the data, represented by an (OWL) ontology, models the domain of interest and hides the complex structure of the underlying data sources. In the ontology, classes and properties are mapped through a declarative specification into views over data expressed in terms of SQL queries. In the virtual approach, the OBDA system first rewrites end-user queries with respect to the ontology, then translates them into SQL queries, and finally delegates the query execution to the SQL engine over relational data sources. OBDA has a strong impact in both scientific and industrial communities.

© Springer International Publishing Switzerland 2016
M. Ortiz and S. Schlobach (Eds.): RR 2016, LNCS 9898, pp. 18–24, 2016.
DOI: 10.1007/978-3-319-45276-0_2

Research in OBDA has grown to maturity and OBDA has become a prominent direction in the development of the Semantic Web. The OWL 2 QL profile of the Web Ontology Language (OWL 2) based on *DL-Lite* [9], a lightweight DL family that enjoys a low complexity of reasoning, has been introduced by the W3C as a standard for OBDA. *Ontop*[1] is a state-of-the-art OBDA engine developed at the Free University of Bozen-Bolzano. Ontop is currently adopted as the core OBDA engine of the EU FP7 Optique project whose goal is to overcome the problem of end-user access to big data [11]. More recently, Ontop has also been integrated in the commercial graph database system *Stardog*[2] to provide support for SPARQL end-user queries.

In many applications data has a temporal dimension, which is important to consider (see, e.g., [13]). In such scenarios, it is reasonable to assume that the concepts of the conceptual OBDA layer have an associated temporal validity periods. If the data, for example, is the stream of wind speed measurements at weather stations, the concept HurricaneForceWind(x, t) can be associated to the data by means of a mapping that extracts the stations and time stamps, where the wind speed exceeded $118\,\text{km/h}$. An ontology designer can then use classical (atemporal) ontology constructors to define new concepts, e.g., "hurricane force wind is a wind". Many studies develop this approach (see [4,7,8,12,14,19] and references therein) and extend the query language of conjunctive or SPARQL queries with the constructors to retrieve temporal information; e.g., "extract stations and timestamps, where hurricane force wind occurred and it also occurred one hour ago". The latter pattern represents the definition of a hurricane (hurricane force wind lasting one hour or longer). The above mentioned approach, in spite of allowing to query for hurricanes, does not allow for defining a concept hurricane that would be very natural in the paradigm of OBDA.

To overcome the limitation of a temporal approach, other studies (see [1,3, 5,15] and references therein) focused on using ontology languages with temporal constructors [2,17] in the setting of OBDA. One can define a hurricane as a new concept by means of temporal operators (e.g., as a conjunction HurricaneForceWind $\wedge$ $\mathbf{X}^-$ HurricaneForceWind), where $\mathbf{X}^-$ is a temporal operator "previous time"). As another example, the concept Blizzard can be defined as an occurrence of Blizzard Condition lasting for more than 3 hours, whereas Blizzard Condition is defined as simultaneous occurrences of Strong Wind, Low Visibility, and Snow (i.e., BlizzardCondition = Strong Wind $\wedge$ Low Visibility $\wedge$ Snow) [18].

The approach that considers atemporal ontologies only is less expressive. It has, however, the advantage that the complexity of the temporal query answering mostly coincides with the complexity of answering usual queries. Therefore, implementations for this setting can be with a reasonable effort reduced to atemporal query answering. With some notable exceptions [1], the complexity of reasoning grows significantly in the approach with temporal ontologies (as compared to reasoning in the underlying ontology languages) [17]. Therefore, it is more

---

[1] http://ontop.inf.unibz.it/.
[2] http://stardog.com/.

challenging to develop a practical query answering system for that setting. As we move towards this goal, we are aware that using more temporal constructors results in higher complexity. Thus, we attempt to allow only those that are necessary for practical use-cases.

The objectives of this ongoing research are: *(a)* to enrich the ontology languages with the operators for building the temporalized concepts that are expressive enough to capture the patterns required by industrial use-cases, *(b)* to implement ontology-mediated query answering with temporalized concepts in the OBDA system Ontop, and *(c)* to perform extensive evaluations using large amounts of real-world data.

As mentioned above, the direction *(a)* has been sufficiently studied. However, more investigation is needed there continuously with respect to new use-cases of temporal OBDA that are being discovered. Regarding the directions *(b)* and *(c)*, to the best of our knowledge, none of the available OBDA implementations take temporal ontologies into account. In this study our aim is to extend the OBDA techniques to support temporal reasoning and implement these techniques in the state-of-the-art framework Ontop.

In the following sections, we identify the research problems and challenges of this study and we propose our methodology. In Sect. 2 we discuss potential applications, in Sect. 3 we explain the main challenges in defining new languages for ontology, mapping, and querying by taking the trade-off between expressivity and efficiency into consideration. In Sect. 4 we analyze the challenges in implementation side of extending the existing system Ontop.

2   Applications and Use Cases

We have already discussed in Sect. 1 how weather concepts such as hurricane and blizzard can be defined using temporal ontologies. Those concepts hold for weather stations (assuming that the data is recorded at them) and time instants. We can then use a role (which can be mapped to an appropriate database) that connects a station with a town, a county, or a state it is located in. Then, one can define, e.g., a (temporal) concept for counties affected by hurricane as "counties which have some station located in them that recorded a hurricane". More interestingly, we can define a concept for cyclone as "states which have four stations located in them such that one of them records southern wind, one northern, one western, and one eastern". (Note that if we have data describing relative position of a station w.r.t. other stations, we can define a cyclone even more precisely.) Another interesting example is a concept for showery counties defined as "counties that have a station that records no precipitation and a station that records precipitation but recorded no precipitation 20 min ago". Use of such and other similar concepts makes sense to detect development of weather in historical or streaming data. A large database of records of weather stations across the US is available through National Weather Service's Mesonet program[3]. It can conveniently be used as a data set to evaluate the performance of our approach.

---

[3] http://mesowest.org.

Another important application of temporal ontologies is analysis of log data of mechanical or electronic devices. For example, if a speed (measured in RpM) of a working engine is continuously recorded in a database, we can extract by means of the mappings such temporal concepts as idle speed, intermediate speed, and running speed. A concept smooth shutdown can then be defined as "idle speed preceded for 15 min by intermediate speed, which is, in its turn, preceded by running speed". On the other hand, rapid shutdown can be defined as "idle speed preceded by occurrence of running speed within 5 min". Another interesting example is a concept consistent vibration defined as "high vibration occurring every 10 s for 1 min". Clearly, using temporal ontologies to conceptually define abnormal situations in performance of devices is a novel and relevant approach to monitoring. We are collaborating with a major industrial company to obtain such data. This company runs several data centers for monitoring thousands of devices related to power generation, including gas and steam turbines, compressors, and generators. Each device is monitored by many sensors of different kinds. These sensors have generated terabytes of data so far. We aim to observe the performance and the scalability of our system over these large amount of real data in collaboration with the researchers in this company.

3   Methodological and Theoretical Challenges

Initially, the most important question to answer is what are the appropriate temporal languages for expressing/formulating ontologies, mappings, and queries in terms of expressivity and efficiency. For both the ontology and the query language level, we have to investigate to which degree the recently proposed temporal ontology languages and query languages satisfy our needs. Below we consider potential challenges in these three areas.

Temporal Extension of the Ontology Language. The first candidate for the role of an ontology language is a Linear Temporal Logic-based Description Logic (DL) proposed in [1], which was shown to have low data complexity (AC^0) for some important fragments. However there are two main reasons that make this logic not perfectly well suited to capture our requirements. On one hand, this logic uses ABoxes with concept assertions of the form $A(a, n)$, where a is an object name and n is a natural number representing a time point. It is often hard to adapt the real-world scenarios to this setting, as neither the timestamps of data records feature fixed periodicity, nor a reasonable atomicity of time in a data source is known a priori. On the other hand, this logic does not provide an explicit way to express metric constraints for temporal concepts (e.g., "hurricane is a strong wind continuing for at least 1 h"). We can overcome the first drawback by using a Halpen-Shoham Interval Logic-based DL proposed in [3, 15], where the ABox concept assertions are assumed to be of the form $A(a, n_1, n_2)$ with n_1, n_2 real or natural numbers indicating a validity interval. This logic was shown be tractable in data complexity too, however, it is even less expressive in terms of the metric constraints, and does not overcome the second drawback. We believe that using ontology languages based on Metric Temporal Logics (MTL) [16] will be needed in our approach. Nothing is known yet neither about the complexity of MTL fragments underlying our temporal constraints, nor about the complexity of reasoning in MTL-based ontology languages.

Temporal Mapping Language. The mapping languages for temporal concepts over log or stream databases is a novel problem that is fundamental to our OBDA approach. In general, if one considers ABoxes with concept assertions of the shape $A(a, n)$, the solution is seemingly easy: a mapping should be an SQL query returning pairs of object names and time stamps. In real-world situations, however, the data may be noisy and, e.g., to detect whether high temperature occurred at a moment of time n, one needs to look at the value of the temperature at several surrounding time moments and take the average. Other approximation functions, such as exponential average, should be considered too, as they are known to be more appropriate for processing certain types of signals. Our aim is to consider both the approximations computable in SQL, as well as other languages for data access.

As mentioned above, in our approach it is more advantageous to consider concept statement of the form $A(a, n_1, n_2)$. Therefore, an SQL query of a mapping should return a pair of time moments, between which, e.g., high temperature occurred. In simple scenarios, where data records are complete for the time stamps, one can use the LEAD function of SQL to compute the n_2 to be "paired" with n_1. In the case when the database is missing values in some fields for some time stamps, computing the temporal concepts may require more elaborate SQL queries ignoring or taking into account (depending on assumptions about a data source) time moments with missing signal values.

Temporal Query Language. Query languages for ontologies over temporal data have been widely considered [4,7,12,14,19], in particular, with SPARQL-inspired syntax [1,20]. In our approach we intend to keep the end-user query language simple by moving the temporal patterns into the ontology level. One important feature that we plan to enable in the queries is the direct use of temporal constants of various granularity, such as 2016, May 2016, afternoon May 5 2016, May 5 2016 11:24, etc. An end-user then can formulate in a natural way queries such as "locations where a blizzard occurred in May 2016", "counties and days when it rained in May 2016", or "engines and minutes where/when consistent vibration occurred in the past hour". We plan to investigate the languages that allow to express such queries.

4 Implementation Challenges

Implementing new forms of mappings and temporal operators of ontologies in Ontop framework will require substantial work. The most reasonable method to store and process the information on validity times of a concept seems to be using the tables (possibly, virtual) representing intervals. In terms of cost efficiency, one of the most challenging tasks in translating temporal operators into SQL queries is computing *coalescing* [6], i.e., the largest time intervals where a concept holds.

For example, in order to compute intervals where hurricane holds, we need to consider a coalescing of the time intervals where hurricane force wind holds. There are various approaches to computing coalescings (e.g., through transitive closure), we are going to investigate what algorithm suits best to our setting.

The other challenging task in translating into SQL is to provide a cost efficient way of performing temporal joins [10]. In the case of the concept for blizzard, in order to get intervals where it holds, one should compute intersections of the intervals where strong wind, low visibility, and snow hold. Given a pair of intervals, one has to consider various relative positions of the first interval w.r.t. the second, in order to find a pair of numbers that represents the intersection. Therefore, a straightforward implementation of the intersection in SQL will result in multiple unions. On the other hand, the CASE operator can help handle those conditions and prevent from making unions, which reduce the performance when a number of joined tables is large. We will also investigate other methods to decrease the temporal join cost by employing SQL cursors. The idea behind using cursors is to perform the temporal join in a merge-sort fashion over the tables that are ordered by starting point of intervals. This approach enables one to apply the temporal join by doing just one iteration of scan over each table that is joined. The drawback of it, however, is that it requires an additional sort step before applying the temporal join.

Acknowledgements. This paper is supported by the EU under the large-scale integrating project (IP) Optique (*Scalable End-user Access to Big Data*), grant agreement n. FP7-318338.

References

1. Artale, A., Kontchakov, R., Kovtunova, A., Ryzhikov, V., Wolter, F., Zakharyaschev, M.: First-order rewritability of temporal ontology-mediated queries. In: Proceedings of IJCAI 2015, pp. 2706–2712. AAAI Press (2015)
2. Artale, A., Kontchakov, R., Ryzhikov, V., Zakharyaschev, M.: A cookbook for temporal conceptual data modelling with description logics. ACM Trans. Comput. Log. **15**(3), 25:1–25:50 (2014)
3. Artale, A., Kontchakov, R., Ryzhikov, V., Zakharyaschev, M.: Tractable interval temporal propositional and description logics. In: Proceedings of AAAI 2015, pp. 1417–1423. AAAI Press (2015)
4. Baader, F., Borgwardt, S., Lippmann, M.: Temporalizing ontology-based data access. In: Bonacina, M.P. (ed.) CADE 2013. LNCS, vol. 7898, pp. 330–344. Springer, Heidelberg (2013)
5. Basulto, V.G., Jung, J., Kontchakov, R.: Temporalized EL ontologies for accessing temporal data: complexity of atomic queries. In: Proceedings of 25th International Joint Conference on Artificial Intelligence (IJCAI-2016). AAAI Press (2016)
6. Böhlen, M.H., Snodgrass, R.T., Soo, M.D.: Coalescing in temporal databases. In: Proceedings of 22th International Conference on Very Large Data Bases (VLDB 1996), pp. 180–191 (1996)
7. Borgwardt, S., Lippmann, M., Thost, V.: Temporal query answering in the description logic DL-Lite. In: Proceedings of FroCoS 2013, pp. 165–180 (2013)

8. Borgwardt, S., Lippmann, M., Thost, V.: Temporalizing rewritable query languages over knowledge bases. J. Web Semant. **33**, 50–70 (2015). Ontology-Based Data Access

9. Calvanese, D., Giacomo, G., Lembo, D., Lenzerini, M., Rosati, R.: Tractable reasoning and efficient query answering in description logics: the DL-Lite family. J. Autom. Reason. **39**(3), 385–429 (2007)

10. Gao, D., Jensen, C.S., Snodgrass, R.T., Soo, M.D.: Join operations in temporal databases. VLDB J. **14**(1), 2–29 (2005)

11. Giese, M., Soylu, A., Vega-Gorgojo, G., Waaler, A., Haase, P., Jiménez-Ruiz, E., Lanti, D., Rezk, M., Xiao, G., Özçep, Ö.L., Rosati, R.: Optique - zooming in on big data access. IEEE Comput. **48**(3), 60–67 (2015)

12. Gutiérrez-Basulto, V., Klarman, S.: Towards a unifying approach to representing and querying temporal data in description logics. In: Krötzsch, M., Straccia, U. (eds.) RR 2012. LNCS, vol. 7497, pp. 90–105. Springer, Heidelberg (2012)

13. Kharlamov, E., et al.: How semantic technologies can enhance data access at siemens energy. In: Mika, P., et al. (eds.) ISWC 2014, Part I. LNCS, vol. 8796, pp. 601–619. Springer, Heidelberg (2014)

14. Klarman, S., Meyer, T.: Querying temporal databases via OWL 2 QL. In: Kontchakov, R., Mugnier, M.-L. (eds.) RR 2014. LNCS, vol. 8741, pp. 92–107. Springer, Heidelberg (2014)

15. Kontchakov, R., Pandolfo, L., Pulina, L., Ryzhikov, V., Zakharyaschev, M.: Temporal and spatial OBDA with many-dimensional Halpern-Shoham logic. In: Proceedings of 25th International Joint Conference on Artificial Intelligence (IJCAI-2016). AAAI Press (2016)

16. Koymans, R.: Specifying real-time properties with metric temporal logic. Real-Time Syst. **2**(4), 255–299 (1990)

17. Lutz, C., Wolter, F., Zakharyaschev, M.: Temporal description logics: a survey. In: Proceedings of TIME 2008, pp. 3–14. IEEE Computer Society (2008)

18. National Oceanic and Atmospheric Administration: National weather service glossary (2009). http://w1.weather.gov/glossary/index.php?letter=b. Accessed 03 Jun 2016

19. Özçep, Ö., Möller, R., Neuenstadt, C., Zheleznyakov, C., Kharlamov, E.: OBDA with temporal and stream-oriented queries: optimization techniques, D5.2. Technical report, FP7-318338, EU (2014)

20. Özçep, Ö.L., Möller, R., Neuenstadt, C.: A stream-temporal query language for ontology based data access. In: Lutz, C., Thielscher, M. (eds.) KI 2014. LNCS, vol. 8736, pp. 183–194. Springer, Heidelberg (2014)

21. Poggi, A., Lembo, D., Calvanese, D., De Giacomo, G., Lenzerini, M., Rosati, R.: Linking data to ontologies. In: Spaccapietra, S. (ed.) Journal on Data Semantics X. LNCS, vol. 4900, pp. 133–173. Springer, Heidelberg (2008)

Semantic Analysis of R2RML Mappings for Ontology-Based Data Access

Cristina Civili[1], Jose Mora[2], Riccardo Rosati[2(✉)], Marco Ruzzi[2], and Valerio Santarelli[2]

[1] School of Informatics, University of Edinburgh, Edinburgh, UK
[2] DIAG, Sapienza Università di Roma, Rome, Italy
rosati@dis.uniroma1.it

Abstract. Ontology-based data access (OBDA) deals with the problem of accessing autonomous data sources through a shared, virtual ontology, and declarative mappings connecting the data sources to the ontology. The W3C standard R2RML allows for mapping relational data sources to RDFS/OWL ontologies. In this paper, we present algorithms for the semantic analysis of R2RML mappings in the OBDA setting, when the ontology is expressed in OWL 2 QL. The focus of such algorithms is to identify the main semantical anomalies (inconsistency and redundancy) of a mapping specification with respect to the ontology and/or the data sources. Such algorithms have been implemented in the mapping analysis tool developed within the Optique European project. We also report on the experiments conducted within the Optique project use cases.

1 Introduction

Ontology-based data access (OBDA) [12] is an approach to the access of multiple, heterogeneous *data sources* through an *ontology* that acts as a shared, abstract model of the data, and a declarative *mapping* that provides the semantic relationship between the data and the ontology.

An *OBDA specification* is the intensional specification of an OBDA setting, i.e., a triple $\langle \mathcal{T}, \mathcal{S}, \mathcal{M} \rangle$ where $\mathcal{T}$ is the ontology, $\mathcal{S}$ is the schema of the data sources and $\mathcal{M}$ is the mapping. In this paper, we focus on the case when $\mathcal{S}$ is a single relational database schema.

Our purpose is to identify algorithms for developing semantic mapping analysis functionalities in an OBDA platform. More precisely, we aim at developing functionalities that help in the construction and maintenance of the OBDA specification. In particular, the present work is motivated by the Optique European project[1] [6], whose aim is to apply OBDA technology in big data scenarios. The issue of creating, debugging and maintaining a mapping specification is a central one in this project, and tools for supporting the design and analysis of mappings are being developed within the project.

Indeed, the specification of the mapping is the most challenging and complex design activity in an OBDA project, since the mapping has to fill the semantic

---

[1] http://www.optique-project.eu/.

© Springer International Publishing Switzerland 2016
M. Ortiz and S. Schlobach (Eds.): RR 2016, LNCS 9898, pp. 25–38, 2016.
DOI: 10.1007/978-3-319-45276-0_3

distance between the ontology and the data sources, which is often very large. So, the declarative assertions constituting the mapping are very complex statements. Moreover, in the Optique use cases, as well as in other practical applications of the OBDA framework (see, e.g., [2]), the number of mapping assertions constituting the mapping is large (hundreds of assertions), and it is extremely difficult to manually handle and debug such a specification.

In this paper we present the mapping analysis component developed within the Optique project, to provide automated support to the specification and debugging of mappings in OBDA. We base our work (Sect. 3) on recent formal notions of *anomalous* mappings in the OBDA context [10,11]: in particular, notions of *inconsistent* and *redundant* mappings, defined both in a *local* and in a *global* version. The local notions refer to single mapping assertions, while the global ones are relative to a whole mapping collection (set of mapping assertions).

We remark that defining an appropriate notion of inconsistency for mappings in OBDA is already challenging, since the "classical" notion of inconsistency is not meaningful. We thus provide a notion of inconsistency for mappings (called *global mapping inconsistency*) that is based on the idea of checking whether the mapping can be "activated" by the data source without creating contradictions with the ontology. On the other hand, a "classical" notion of redundancy (that is, the one that naturally follows from the semantics of an OBDA system) appears appropriate for our purposes.

This formal framework allows us (Sect. 4) to attack the problem of defining concrete algorithms for semantic mapping analysis in OBDA. However, differently from [11], here we consider the W3C standard R2RML [5] as the mapping language. Such a language allows for expressing arbitrary SQL queries over the database source. This immediately makes almost every significant semantic check over R2RML mappings undecidable, independently of the ontology language (or equivalently, even if the ontology is empty). Nevertheless, we are able to define approximated techniques for semantic mapping analysis based on: (i) the translation of SQL into first-order logic; (ii) the usage of a first-order theorem prover to solve reasoning problems that encode the additional expressiveness of R2RML with respect to GAV and GLAV.

Finally, in Sect. 5 we present the experimental results obtained by our mapping analysis algorithms in the Optique project use cases.

2   Preliminaries

In the following, we assume to have four pairwise disjoint, countably infinite alphabets: an alphabet Γ_T of ontology predicates, an alphabet Γ_S of source schema predicates, an alphabet Γ_C of constants, and an alphabet Γ_F of functions.

Source schemas. A source schema S is a relational schema containing relations in Γ_S, possibly equipped with integrity constraints (ICs). A *legal instance D for* S is a database for S (i.e., a finite set of ground atoms over S and the constants in Γ_C) that satisfies the ICs of S. We denote by $Const(D)$ the set of constants occurring in D.

We consider integrity constraints corresponding to first-order sentences. Given a source schema $\mathcal{S}$, we denote by $\Psi(\mathcal{S})$ the first-order sentence constituted by the conjunction of the sentences corresponding to its integrity constraints.

Given a first-order sentence α, we write $\mathcal{S} \models \alpha$ if for each database D legal for $\mathcal{S}$, $\mathcal{I}_D \models \alpha$, where $\mathcal{I}_D$ is the interpretation induced by D.

We call *simple schema* a source schema without ICs. We adopt standard notions for first-order (FO) queries and conjunctive queries (CQs) over relational schemas [1]. By a FO query over a source schema $\mathcal{S}$ we mean a FO query over the alphabet of $\mathcal{S}$. With $\phi(\boldsymbol{x})$ we denote a FO query with free variables $\boldsymbol{x}$. The number of variables in $\boldsymbol{x}$ is the arity of the query. A Boolean FO query is a FO query without free variables. Given a FO q over $\mathcal{S}$ and a legal instance D for $\mathcal{S}$, $eval(q, D)$ denotes the evaluation of q over D. In what follows, we will always denote a source schema with $\mathcal{S}$.

Ontologies. We consider ontologies expressed in the description logic *DL-Lite$_R$* [4], the logic underlying the OWL 2 QL standard profile.[2] In particular, a *DL-Lite$_R$* ontology $\mathcal{O}$ is a pair $\langle \mathcal{T}, \mathcal{A} \rangle$, where $\mathcal{T}$ is the TBox and $\mathcal{A}$ is the ABox. In what follows, $\mathcal{O}$, $\mathcal{T}$, and $\mathcal{A}$, respectively, will always have the same meaning. As in the W3C standard OWL, we do not interpret ontologies under the Unique Name Assumption. We denote with $Models(\mathcal{O})$ the set of models of $\mathcal{O}$, and with $\mathcal{O} \models \alpha$ the fact that $\mathcal{O}$ entails a sentence α. Also, by *ontology inconsistency* we mean the task of deciding whether $Models(\mathcal{O}) = \emptyset$, and by *instance checking* the task of deciding whether $\mathcal{O} \models \beta$, where β is a ground atom. By *CQs over $\mathcal{O}$* we mean CQs over the alphabet of the TBox of $\mathcal{O}$, and by *CQ entailment* the task of checking whether $\mathcal{O} \models q$, where q is a Boolean CQ.

Mappings. A *mapping assertion* m from a source schema $\mathcal{S}$ to a TBox $\mathcal{T}$ has the form

$$\phi(\boldsymbol{x}) \rightsquigarrow \psi(\boldsymbol{x}) \tag{1}$$

where $\phi(\boldsymbol{x})$ is a function-free first-order query with free variables $\boldsymbol{x}$ (and, possibly, existentially quantified variables) over the predicates of $\mathcal{S}$, and $\psi(\boldsymbol{x})$ is a conjunctive query with function symbols, i.e., a conjunction of atoms whose predicates are concepts and roles from $\mathcal{T}$ and whose arguments may be variables from $\boldsymbol{x}$, constants, or terms of the form $f(t_1, \ldots, t_n)$ where $n \geq 1$, $f \in \Gamma_{\mathcal{F}}$ and every t_i is either a variable from $\boldsymbol{x}$ or a constant. The free variables $\boldsymbol{x}$ are called the *frontier variables* of m, and denoted by $FR(m)$. Moreover, $\phi(\boldsymbol{x})$ is called the *body of m* (denoted by $body(m)$), and $\psi(\boldsymbol{x})$ is called the *head of m* (denoted by $head(m)$). The number of variables in $\boldsymbol{x}$ is the *arity* of the mapping assertion. A mapping $\mathcal{M}$ from $\mathcal{S}$ to $\mathcal{T}$ is a finite set of mapping assertions from $\mathcal{S}$ to $\mathcal{T}$. Hereinafter $\mathcal{M}$ will always denote a mapping.

The above defined mapping language is the one typically considered in OBDA [3,12], and captures almost all the R2RML W3C standard mapping language [5].

---

[2] http://www.w3.org/TR/owl2-profiles/.

We say that a mapping assertion m *is active on a source instance D* if $eval(body(m), D)$ is a non-empty set of tuples of constants. A mapping $\mathcal{M}$ is active on D if all its mapping assertions $m \in \mathcal{M}$ are active on D.

Without loss of generality, we assume that different mapping assertions use different variable symbols. A *freeze* of a set of atoms Γ is a set of ground atoms obtained from Γ by replacing every variable with a *fresh* distinct constant. In this paper, the freeze is always used in the context of a mapping $\mathcal{M}$, so it suffices to assume that fresh constants do not appear in $\mathcal{M}$. Different freezes of the same set of atoms are equal up to renaming of constants. Thus, in the following we assume, without loss of generality, that the freeze of a set of atoms Γ is unique and is obtained by replacing each variable occurrence x with a fresh constant c_x, and we denote it by $Freeze(\Gamma)$.

Given a mapping assertion m of arity n and an n-tuple of constants t, we denote by $m(t)$ the mapping assertion obtained by replacing $FR(m)$ in m with the constants in t.

OBDA Specifications. An OBDA specification is a triple $\mathcal{J} = \langle \mathcal{T}, \mathcal{S}, \mathcal{M} \rangle$. The semantics of $\mathcal{J}$ is given with respect to a database instance D legal for $\mathcal{S}$: a model for $\mathcal{J}$ w.r.t. D is a FOL interpretation $\mathcal{I}$ over the alphabet $\Gamma_{\mathcal{T}} \cup \Gamma_{\mathcal{C}} \cup \Gamma_{\mathcal{F}}$ that satisfies both $\mathcal{T}$ and $\mathcal{M}$. Formally, we say that $\mathcal{I}$ satisfies the mapping $\mathcal{M}$ if for each assertion $m \in \mathcal{M}$ and each tuple of constants t such that $t \in eval(body(m), D)$ we have that $\mathcal{I} \models head(m(t))$. The set of models of $\mathcal{J}$ w.r.t. D is denoted with $Models(\mathcal{J}, D)$. Also, we use $(\mathcal{J}, D)$ to denote $\mathcal{J}$ with source instance D. We say that $(\mathcal{J}, D)$ *is inconsistent* if $Models(\mathcal{J}, D) = \emptyset$, and denote with $(\mathcal{J}, D) \models \alpha$ the entailment of a sentence α by $(\mathcal{J}, D)$.

Example 1. As an example of an OBDA specification, we consider a source schema $\mathcal{S}$ where the **plants** relation contains data on extraction facilities, while the **eZones** relation contains data on the areas used for oil and gas extraction. Below, the underlined attributes represent the keys of the relations.

$$\text{plants}(\underline{\text{id\_pl}}, \text{pl\_typ}, \text{id\_zn}) \qquad \text{eZones}(\underline{\text{id\_zn}}, \text{zn\_typ})$$

The formula $\Psi(\mathcal{S})$ expressing the source schema $\mathcal{S}$ is the following:

$$(\forall x, y, z, y', z'. \text{plants}(x, y, z) \wedge \text{plants}(x, y', z') \rightarrow (y = y' \wedge z = z')) \wedge$$
$$(\forall x, y, y'. \text{eZones}(x, y) \wedge \text{eZones}(x, y') \rightarrow y = y')$$

The following *DL-Lite$_R$* TBox models a very small portion of the domain of oil and gas production extracted from an ontology developed within the Optique project. In particular, the TBox focuses on the facilities (concept Facility) used in the oil and gas extraction and on the geographical areas (concept Area) in which they are located (role locatedIn). A marine area (concept MarArea) is a subconcept of the concept Area.

$\mathcal{T} = \{$ Platform $\sqsubseteq$ Facility,   MarArea $\sqsubseteq$ Area, $\exists$locatedIn $\sqsubseteq$ Facility,
         $\exists$locatedIn$^-$ $\sqsubseteq$ Area   Facility $\sqcap$ Area $\sqsubseteq \bot \}$

The following is an example of a mapping $\mathcal{M}$ from $\mathcal{S}$ to $\mathcal{T}$:

$m_1 :\ (\exists y.\, \texttt{plants}(x, y, z))\ \rightsquigarrow\ \mathsf{Facility}(f(x)) \wedge \mathsf{locatedIn}(f(x), z)$
$m_2 :\ \texttt{plants}(x', \text{'pl'}, y')\ \rightsquigarrow\ \mathsf{Platform}(p(x'))$
$m_3 :\ \texttt{eZones}(z', \text{'mz'})\ \rightsquigarrow\ \mathsf{MarArea}(m(z')).$

□

3   Formal Notions of Mapping Anomalies

In this section we recall the formal framework of [10,11] that constitutes the basis of the mapping analysis functionalities that will be studied in the next section. We first deal with mapping consistency, then we turn our attention to mapping redundancy and subsumption. All the definitions of this section are taken from [11], with the exception of Definition 2.

3.1   Mapping Inconsistency

We start by providing a "global" notion of inconsistency, that is, inconsistency relative to a whole mapping specification.

Definition 1 (Global Mapping Inconsistency). *Let $\mathcal{J} = \langle \mathcal{T}, \mathcal{S}, \mathcal{M} \rangle$ be an OBDA specification. We say that $\mathcal{M}$ is globally inconsistent for $\langle \mathcal{T}, \mathcal{S} \rangle$ if there does not exist a source instance D legal for $\mathcal{S}$ such that $\mathcal{M}$ is active on D and $Models(\mathcal{J}, D) \neq \emptyset$.*

Intuitively, if a mapping is globally inconsistent, then it is not possible to simultaneously activate all its mapping assertions without causing inconsistency of the whole specification. This is certainly an anomalous situation, as shown by the following example.

Example 2. Let $\mathcal{J} = \langle \mathcal{T}, \mathcal{S}, \mathcal{M} \rangle$ be an OBDA specification where $\mathcal{T}$ and $\mathcal{S}$ are as in Example 1. Suppose that the mapping $\mathcal{M}$ contains the following mapping assertions:

$m_1 :\ (\exists y, z.\, \texttt{plants}(x, y, z))\ \rightsquigarrow\ \mathsf{Area}(x)$
$m_2 :\ \texttt{plants}(x', \text{'pl'}, z')\ \rightsquigarrow\ \mathsf{Platform}(x') \wedge \mathsf{locatedIn}(x', z')$

It is easy to see that $\mathcal{M}$ is globally inconsistent for $\langle \mathcal{T}, \mathcal{S} \rangle$, because $\mathcal{T} \models \mathsf{Platform} \sqcap \mathsf{Area} \sqsubseteq \bot$ and every activation of m_2 also activates m_1, thus implying $\mathsf{Platform}(x)$ and $\mathsf{Area}(x)$ for the same individual x.   □

Then, we provide a novel notion of *strong* local mapping inconsistency.[3]

Definition 2 (Strong Local Mapping Inconsistency). *Let $\mathcal{T}$ be a TBox and let $\mathcal{S}$ be a source schema. We say that a mapping assertion m is strongly locally inconsistent for $\langle \mathcal{T}, \mathcal{S} \rangle$ if there does not exist a source instance D legal for $\mathcal{S}$ such that $\{m\}$ is active on D and $Models(\langle \mathcal{T}, \mathcal{S}, \{m\} \rangle, D) \neq \emptyset$.*

---

[3] This notion of *strong local inconsistency* is slightly different from the notion of *local inconsistency* presented in [11]: in particular, it can be shown that strong local consistency implies local consistency, while the converse in general does not hold.

In practice, the notion of strong local inconsistency corresponds to check the inconsistency of a single mapping assertion with respect to $\langle \mathcal{T}, \mathcal{S} \rangle$.

Note that the strong local mapping inconsistency of $m \in \mathcal{M}$ for $\langle \mathcal{T}, \mathcal{S} \rangle$ implies the global mapping inconsistency of $\mathcal{M}$ for $\langle \mathcal{T}, \mathcal{S} \rangle$. On the other hand, a mapping $\mathcal{M}$ that is globally inconsistent for some $\langle \mathcal{T}, \mathcal{S} \rangle$ may not contain any mapping assertion m that is inconsistent for $\langle \mathcal{T}, \mathcal{S} \rangle$. That is, the strong local inconsistency of a mapping assertion is a sufficient but not necessary condition for global inconsistency.

3.2   Mapping Redundancy

We now deal with mapping redundancy. First, given an ODBA specification $\mathcal{J} = \langle \mathcal{T}, \mathcal{S}, \mathcal{M} \rangle$ where $\mathcal{M} = \{m\}$, we consider a mapping assertion m' to be redundant for m, if adding m' to $\mathcal{M}$ produces a specification equivalent to $\mathcal{J}$. This is formalized below.

Definition 3 (Local Mapping Redundancy). *Let $\mathcal{T}$ be a TBox, let $\mathcal{S}$ be a source schema, and let m, m' be mapping assertions of the same arity from $\mathcal{S}$ to $\mathcal{T}$. We say that m' is redundant for m under $\langle \mathcal{T}, \mathcal{S} \rangle$ if, for every source instance D that is legal for $\mathcal{S}$, $Models(\langle \mathcal{T}, \mathcal{S}, \{m\} \rangle, D) = Models(\langle \mathcal{T}, \mathcal{S}, \{m, m'\} \rangle, D)$.*

Example 3. Let $\mathcal{T}$ and $\mathcal{S}$ be as in Example 1. Consider the following mapping assertions:

$m_1 :$ $\texttt{plants}(x, \text{'pl'}, z)$ $\quad \rightsquigarrow \quad$ $\mathsf{locatedIn}(x, z)$
$m_2 :$ $(\exists y. \texttt{plants}(x, y, z))$ $\rightsquigarrow$ $\mathsf{Facility}(x) \wedge \mathsf{locatedIn}(x, z)$

It is easy to see that the m_1 mapping assertion is locally redundant for m_2 under $\langle \mathcal{T}, \mathcal{S} \rangle$.   □

Then, we define a more general, global notion of mapping redundancy which is relative to a whole mapping specification.

Definition 4 (Global Mapping Redundancy). *Let $\mathcal{J} = \langle \mathcal{T}, \mathcal{S}, \mathcal{M} \rangle$ be an OBDA specification and let $\mathcal{M}'$ be a mapping from $\mathcal{S}$ to $\mathcal{T}$. We say that $\mathcal{M}'$ is globally redundant for $\mathcal{J}$ if, for every source instance D that is legal for $\mathcal{S}$, $Models(\langle \mathcal{T}, \mathcal{S}, \mathcal{M} \rangle, D) = Models(\langle \mathcal{T}, \mathcal{S}, \mathcal{M} \cup \mathcal{M}' \rangle, D)$.*

Example 4. Let $\langle \mathcal{T}, \mathcal{S}, \mathcal{M} \rangle$ be an OBDA specification, where $\mathcal{T}$ and $\mathcal{S}$ are as in Example 1, and $\mathcal{M}$ is as follows:

$m_1 :$ $(\exists y. \texttt{plants}(x, y, z) \wedge \texttt{eZones}(z, \text{'mz'}))$ $\quad \rightsquigarrow \quad$ $\mathsf{locatedIn}(x, z)$
$m_2 :$ $\texttt{eZones}(x', \text{'mz'})$ $\quad \rightsquigarrow \quad$ $\mathsf{MarArea}(x')$
$m_3 :$ $\texttt{plants}(y', \text{'pl'}, z') \wedge \texttt{eZones}(z', \text{'mz'})$ $\quad \rightsquigarrow \quad$ $\mathsf{locatedIn}(y', z') \wedge \mathsf{Area}(z')$

Then, $\{m_3\}$ is globally redundant for $\langle \mathcal{T}, \mathcal{S}, \{m_1, m_2\} \rangle$.   □

Notice that global redundancy of a mapping $\mathcal{M}'$ for a mapping $\mathcal{M}$ under $\langle \mathcal{T}, \mathcal{S} \rangle$ does not imply that there exists an assertion m' in $\mathcal{M}'$ and an assertion m in $\mathcal{M}$ such that m' is redundant for m under $\langle \mathcal{T}, \mathcal{S} \rangle$, as shown below.

Example 5. Consider the ontology $\mathcal{T} = \{A_1 \sqsubseteq A, B_1 \sqsubseteq B\}$, the source schema composed by the only unary predicate Q, and the following mapping assertions:

$$\begin{aligned}
m_1 &: Q(X) \rightsquigarrow A_1(X) \\
m_2 &: Q(X) \rightsquigarrow B_1(X) \\
m_3 &: Q(X) \rightsquigarrow A(X) \wedge B(X)
\end{aligned}$$

Then, $\mathcal{M}' = \{m_3\}$ is globally redundant for $\langle \mathcal{T}, \mathcal{S}, \{m_1, m_2\}\rangle$, but m_3 is not locally redundant under $\langle \mathcal{T}, \mathcal{S}\rangle$ for any mapping assertion in $\mathcal{M}$.   □

Conversely, it is easy to see that if a mapping $\mathcal{M}'$ contains only assertions that, taken one by one, are redundant under $\langle \mathcal{T}, \mathcal{S}\rangle$ for some assertion contained in a mapping $\mathcal{M}$, then $\mathcal{M}'$ is globally redundant for $\mathcal{M}$ under $\langle \mathcal{T}, \mathcal{S}\rangle$.

Finally, we observe that local mapping redundancy is a special case of global mapping redundancy in which the mapping $\mathcal{M}$ and $\mathcal{M}'$ are both singleton.

4   Algorithms for the Optique System

The techniques for mapping analysis implemented within the Optique system are based on the construction of a matrix of ABox assertions. More precisely, given a mapping $\mathcal{M}$ relative to a source schema $\mathcal{S}$, we define the *ABox matrix for $\mathcal{M}$ under source schema $\mathcal{S}$*, and denote it by $AM(\mathcal{M}, \mathcal{S})$. We will then show that such an ABox matrix can be used to reduce all the mapping consistency and redundancy tasks defined in the previous section to standard DL ontology reasoning tasks (ontology consistency and instance checking).

First, we introduce some preliminary definitions.

The *(partial) grounding g* of the frontier variables of a mapping assertion m is a partial function from $FR(m)$ to a set of constants.

Given two groundings g_1 and g_2 for m, if g_1 is equal to g_2 on all the variables mapped by g_2 and there exists $x \in FR(m)$ that is mapped by g_1 and is not mapped by g_2, then we say that g_1 is *preferred* to g_2 for m.

Given a mapping assertion m, we denote by *Freeze$_{FR}(m)$* the mapping assertion obtained from m by freezing of the frontier variables of m: more precisely, in *Freeze$_{FR}(m)$* every occurrence of the frontier variable x is replaced by the constant c_x (w.l.o.g., we assume that different mapping assertions use different variable symbols, and that none of the c_x's appears in $\mathcal{M}$).

Finally, given a mapping assertion m, we denote by *Const$_{FR}(m)$* the set of constants $\{c_x \mid x \in FR(m)\}$.

4.1   The Algorithm BuildABoxMatrix

We are now ready to present the algorithm that builds the ABox matrix $AM(\mathcal{M}, \mathcal{S})$:

Algorithm. BuildABoxMatrix($\mathcal{M}, \mathcal{S}$)
Input: mapping $\mathcal{M} = \{m_1, \ldots, m_n\}$, source schema $\mathcal{S}$
Output: $AM(\mathcal{M}, \mathcal{S})$
begin
  for i:=1 to n **do**
    for j:=1 to n **do begin**
      $M[i,j] = \emptyset$;
      for each grounding $g : FR(m_j) \rightarrow Const_{FR}(m_i)$ **such that**
        (i) $\Psi(\mathcal{S}) \models body(Freeze_{FR}(m_j)) \rightarrow g(body(m_i))$
      and
        (ii) there exists no grounding $g' : FR(m_j) \rightarrow Const_{FR}(m_i)$
          such that $\Psi(\mathcal{S}) \models body(Freeze_{FR}(m_j)) \rightarrow g'(body(m_i))$
          and g' is preferred to g for m
      do $M[i,j] := M[i,j] \cup Freeze(head(g(m_i)))$
    end;
    return M
end

Informally, the ABox matrix M computed by the above algorithm BuildABoxMatrix($\mathcal{M}, \mathcal{S}$) is such that every cell $M[i,j]$ represents, through ABox assertions, *how m_i is activated by m_j*: every cell $M[i,j]$ is a set of ABox assertions that represent (using "frozen" individual names) the concept and role instances retrieved by the mapping assertion m_i when the mapping assertion m_j is active on any database instance D. More precisely, if (a projection of) the query in the body of assertion m_j is contained in (a projection of) the query in the body of assertion m_i (condition (i) in the algorithm), then any activation of m_j implies the activation of m_i: this is a crucial property both for mapping inconsistency and for mapping redundancy. The ABox matrix represents such semantic dependencies through ABox assertions that use the same individuals.

Example 6. Consider the following mapping $\mathcal{M}$ (on a simple source schema $\mathcal{S}$):

$$m_1 : (\exists z. T_1(x,y,z) \wedge T_2(z,y) \wedge T_3(y,x)) \rightsquigarrow C(x) \wedge R(x,y)$$
$$m_2 : (\exists y', z'. T_1(x',y',z')) \rightsquigarrow D(x')$$
$$m_3 : (\exists z''. T_2(z'',y'') \wedge T_3(y'',x'')) \rightsquigarrow S(x'',y'')$$

The ABox matrix M returned by the algorithm BuildABoxMatrix($\mathcal{M}, \mathcal{S}$) is as follows:

	1	2	3
1	$\{C(c_x), R(c_x, c_y)\}$		
2	$\{D(c_x)\}$	$\{D(c_{x'})\}$	
3	$\{S(c_x, c_y)\}$		$\{S(c_{x''}, c_{y''})\}$

In particular, the presence of the $D(c_x)$ in $M[2,1]$ encodes the fact that any activation of the mapping assertion m_1 implies that the mapping assertion m_2 is also activated (because the body query of m_1 is contained into the body query of m_2). Similarly, the presence of the $S(c_x, c_y)$ in $M[3, 1]$ encodes the fact that any activation of the mapping assertion m_1 also implies the activation of mapping m_3 (because the body query of m_1 is contained into the body query of m_3).

4.2 Limits of the Algorithm

The algorithm BuildABoxMatrix and its implementation have two main limitations.

First, both check (i) and check (ii) in the above algorithm require to decide the validity of an arbitrary first-order sentence. This of course is an undecidable problem, so the above checks can only be approximated by our implementation of the algorithm. In particular, we have used the E theorem prover[4] to solve the above mentioned validity checks, using a time-out (which we configured in a range from 5 to 30 s) for every task.

In the case when no answer is provided within the time-out, our implementation assumes a "no" answer (i.e., no dependency between the two body queries). Therefore, some dependency between mapping assertions may be not represented by the ABox matrix returned by the algorithm. We believe that, given the goal of providing semantic support in the debugging phase of the mapping, this choice is better than assuming a "yes" answer in the cases not decided by the E prover, since in this case "false positives" would be produced then by the inconsistency and redundancy checks that make use of the ABox matrix.

Second, while the ABox matrix "materializes" (through concept and role instances) in a correct way the semantic relationship between two mapping assertions, there are more complex dependencies that are not captured by the matrix. For instance, consider the following mapping $\mathcal{M}$ (on a simple source schema $\mathcal{S}$):

$$m_1 : T_1(x, y) \rightsquigarrow R(x, y)$$
$$m_2 : T_2(x', y') \rightsquigarrow S(x', y')$$
$$m_3 : T_1(x'', y'') \wedge T_2(z'', w'') \rightsquigarrow P(x'', w'')$$

Here, the activation of a single mapping assertion does not imply the activation of any other mapping assertion. However, it is immediate to see that the activation of both m_1 and m_2 implies the activation of the assertion m_3. This is not captured by the ABox matrix, which only considers dependencies between single mapping assertions.

To overcome such an incompleteness, the algorithm should consider simultaneous activations of arbitrary subsets of mapping assertions: however, this would have a dramatic impact on the performance of the algorithm, since it would require an exponential number of iterations rather than a quadratic one.

We believe that such an incompleteness is in practice not problematic, since in real cases the probability of dealing with situations in which the analysis of simultaneous activations of multiple mapping assertions is required is very low.

---

[4] http://wwwlehre.dhbw-stuttgart.de/~sschulz/E/E.html.

Therefore, due to both the above described limitations, the ABox matrix actually represents only a partial picture of the semantic dependencies among the mapping assertions. Despite such a limitation, we can still provide a significant semantic analysis of mappings.

4.3   Checking Mapping Inconsistency and Redundancy Through the ABox Matrix

We now show how the ABox matrix can be used to solve the mapping consistency and redundancy problems introduced in the previous section.

Strong Local Consistency. Let $\mathcal{T}$ be a TBox, let $\mathcal{S}$ be a source schema, let $\mathcal{M}$ be a mapping, let $|\mathcal{M}| = n$, let M be the matrix returned by the algorithm BuildABoxMatrix($\mathcal{M}, \mathcal{S}$), let i be an integer such that $1 \leq i \leq n$, and let $\mathcal{A}$ be the ABox defined as follows:

$$\mathcal{A} = M[i, i]$$

If the ontology $\langle \mathcal{T}, \mathcal{A} \rangle$ is inconsistent, then m_i is strongly inconsistent for $\langle \mathcal{T}, \mathcal{S} \rangle$.

Global Consistency. Let $\mathcal{T}$ be a TBox, let $\mathcal{S}$ be a source schema, let $\mathcal{M}$ be a mapping, let $|\mathcal{M}| = n$, let M be the matrix returned by the algorithm BuildABoxMatrix($\mathcal{M}, \mathcal{S}$) and let $\mathcal{A}$ be the ABox defined as follows:

$$\mathcal{A} = \bigcup_{i=1}^{n} \bigcup_{j=1}^{n} M[i, j]$$

If the ontology $\langle \mathcal{T}, \mathcal{A} \rangle$ is inconsistent, then $\mathcal{M}$ is globally inconsistent for $\langle \mathcal{T}, \mathcal{S} \rangle$.

Local Redundancy. Let $\mathcal{T}$ be a TBox, let $\mathcal{S}$ be a source schema, let $\mathcal{M}$ be a mapping, let $|\mathcal{M}| = n$, let M be the matrix returned by the algorithm BuildABoxMatrix($\mathcal{M}, \mathcal{S}$), and let i, j be integers such that $1 \leq i \leq n$ and $1 \leq j \leq n$. Now let $\mathcal{A}$ be the ABox defined as follows:

$$\mathcal{A} = M[j, i]$$

If $\langle \mathcal{T}, \mathcal{A} \rangle \models M[i, i]$, then m_i is redundant for m_j under $\langle \mathcal{T}, \mathcal{S} \rangle$.

Global Redundancy. Let $\mathcal{T}$ be a TBox, let $\mathcal{S}$ be a source schema, let $\mathcal{M}$ be a mapping, let $|\mathcal{M}| = n$, let M be the matrix returned by the algorithm BuildABoxMatrix($\mathcal{M}, \mathcal{S}$), let i be an integer such that $1 \leq i \leq n$, and let $\mathcal{M}' = \mathcal{M} \setminus \{m_i\}$. Now let $\mathcal{A}$ be the ABox defined as follows:

$$\mathcal{A} = \bigcup_{j \in \{1, \dots, i-1, i+1, \dots, n\}} M[j, i]$$

If $\langle \mathcal{T}, \mathcal{A} \rangle \models M[i, i]$, then m_i is globally redundant for $\langle \mathcal{T}, \mathcal{S}, \mathcal{M}' \rangle$.

Using the above properties, we have implemented algorithms based on the ABox matrix for both local and global mapping inconsistency and for both local and global mapping redundancy.

5   Experiments

The algorithms presented in this paper have been implemented as a novel mapping analysis component within the Optique European project, and, as all other components and APIs developed by the project partners, integrated on the Optique platform through the Information Workbench (IWB) [8]. IWB is a semantic data management and integration platform which provides a shared triple store for managing OBDA system assets, i.e., ontologies, mappings, database metadata, and queries.

Implementation of the mapping analysis component consists both in the addition of new features to the IWB mapping component and in integration with already existing mapping editing features. Namely, the latter allows a combination of mapping editing and analysis through automatic execution of syntactic checks on new or edited mapping rules. The former instead enriches the mapping component with the following capabilities.

1. Syntactic, local and global checks (for both inconsistency and redundancy) on any mapping available in the Optique IWB repository.
2. *Explanation* of the mapping analysis results. Noticeably, for inconsistency checks, the explanation or, potentially, the explanations in the case of global inconsistency, are provided in terms of the combination of the set of TBox axioms and the single ABox axiom, among the ones generated by algorithm makeABox, that together determine an inconsistency. This set of axioms is produced by using the HermiT reasoner [7] and the OWL API BlackBoxGenerator and HSTExplanationGenerator classes. Furthermore, *provenance* of the ABox axiom in each inconsistency explanation is provided. In other words, for each such axiom, the set of mapping assertions whose activation in algorithm makeAbox concurs either directly or indirectly to produce the axiom is returned.
3. Materialization of the mapping analysis results in the shared Optique repository hosted by the IWB platform. All mapping analysis results are translated into RDF triples and stored in the repository for future querying. In case of addition, deletion, or modification of one or more mapping assertions in a mapping, mapping analysis is automatically reset by the system, and all mapping analysis results are deleted from the repository.

The IWB provides the user with a *Semantic Wiki*, whose template pages are automatically instantiated for resources of some fixed type. The wiki features a table-centric interface, in which information is provided mostly in table form. Such tables are populated by the RDF resources in the IWB's repository that are the result of pre-defined structured SPARQL queries.

The interface of the mapping analysis component inside IWB is provided through extensions of the TriplesMapCollection and MappingCollection templates, which show, respectively, the available mappings in the repository, and information about a single mapping. A "Mapping Analysis Report" section has been added to the TriplesMapCollection template, showing, for each mapping in the repository, the ontology referenced by the mappings, the status of the mapping analysis, i.e., whether it has been performed or not, and, if so, whether or not there are local or global inconsistencies, and if the explanations have been computed. Instead, the MappingCollection template has been extended with an "Analysis Results" section, detailing the anomalies identified for each performed check: for each global inconsistency, a reference to its explanations; for each syntactically incorrect or locally inconsistent mapping assertion in the mapping, the reference to the mapping and a message detailing the anomaly; for each local subsumption, the subsumer and subsumee mapping assertions and a message detailing the type of subsumption, e.g., a head or body subsumption; finally, for global redundancies, the redundant mapping assertion and a message explaining the redundancy. Furthermore, custom templates have been produced for mapping inconsistency explanations and for explanation provenance, showing, for each, the relevant information described above, i.e., for explanations, the set of ontology axioms involved in the explanation and a reference to the provenance of the ABox axiom, and for provenance, the mappings responsible for producing the axiom.

The performance of the mapping analysis component was evaluated on one of the two large-scale use cases of the Optique project from the energy sector, namely the Statoil use case [9].

In this scenario, expert geologists develop stratigraphic models of unexplored areas on the basis of data acquired from previous operations at nearby geographical locations through advanced visual analytics tools that access more than one thousand terabytes of data. The ontology developed for the Statoil use case describes wellbores that are drilled for the extraction of natural resources such as gas or oil, and stratigraphic columns of rock layers in the geographical areas interested by these wellbores. It also describes the different kinds of measurements that can be performed in wellbores. The ontology consists of about 150 concepts and 100 roles and attributes.

The Statoil use case features two different data sources: the Exploration and Production Data Store (EPDS), and the NPD FactPages (NPD FP). EPDS is Statoil's corporate data store for exploration and production data and their own interpretations of this data, while NPD FP is a publicly available dataset that is published and maintained by the Norwegian authorities, containing reference data for many aspects of the Norwegian petroleum industry, and is often used as a data source by geologists in combination with EPDS. The mapping used in the mapping analysis evaluation relates to the EPDS data store, which currently has about 3,000 tables with about 37,000 columns, and contains about 700 gigabytes of data.

The evaluation was performed on a version of the EPDS mappings from February 2015, which is formed by 81 mapping assertions. The syntactic, consistency and redundancy tests were conducted incrementally, to account for the fact that a local inconsistency of a mapping assertion entails the global inconsistency of the mapping (hence, to find a global inconsistency that does not depend on local inconsistencies, there must be none of the latter in the mapping), and

Table 1. Results of the evaluation of the mapping analysis on the February 2015 version of the Statoil mappings for the EPDS data source.

Mapping check	Anomalies found	Time (sec)
Syntactic	5	.12
Local consistency	7	2.32
Global consistency	1	26.76
Local redundancy	2	55.33
Global redundancy	3	84.28

that redundancy checks must be performed on a consistent ontology. Therefore, evaluation was performed in the following steps.

1. Identification of syntactically incorrect and locally inconsistent mapping assertions.
2. Removal of locally inconsistent mapping assertions.
3. Identification of global inconsistencies in the mapping and production of explanations for each global inconsistency.
4. Removal of the mapping assertions, highlighted in the explanations, responsible for the global inconsistencies.
5. Identification of local and global redundancies.

The results of the evaluation are provided in Table 1. Syntactic correctness, local consistency and global redundancy were checked for each mapping assertion in the mapping, local redundancy was checked for each pair of mapping assertions in the mapping, and global consistency was checked for the whole mapping. In the case of global consistency, a value of "1" in the table indicates that the mapping was globally inconsistent, and for such an inconsistency 18 different explanations were produced. All execution times are expressed in seconds, and the complete mapping analysis procedure took roughly 3 min.

6   Conclusions

In this paper we have presented algorithm for the semantic analysis of R2RML mappings in the context of ontology-based data access. In particular, we have focused on the OWL 2 QL ontology language. We have also presented an experimental evaluation of our algorithms in the Optique project use cases.

We believe that supporting the design and maintenance of OBDA specifications, and the semantic analysis of mappings in particular, is a crucial aspect towards the successful depolyment of the OBDA technology in the real world. Within the Optique system, we are currently further expanding the mapping analysis component. First, we are developing new functionalities that make use of the ABox matrix presented in this paper. In particular, we are implementing a mapping evolution and repair functionality. In addition, we are defining a new *instance-level* mapping debugging technique, which expolits information about wrong or missing concept and role instances to identify the subset of mapping assertions that need to be repaired.

Acknowledgments. This research has been partially supported by the EU under FP7 project Optique (grant n. FP7-318338).

References

1. Abiteboul, S., Hull, R., Vianu, V.: Foundations of Databases. Addison Wesley Publ. Co., Reading (1995)
2. Antonioli, N., Castanò, F., Civili, C., Coletta, S., Grossi, S., Lembo, D., Lenzerini, M., Poggi, A., Savo, D.F., Virardi, E.: Ontology-based data access: the experience at the Italian Department of treasury. In: Proceedings of the Industrial Track of the 25th International Conference on Advanced Information Systems Engineering (CAiSE), vol. 1017 of CEUR Electronic Workshop Proceedings, pp. 9–16 (2013). http://ceur-ws.org/
3. Bagosi, T., et al.: The ontop framework for ontology based data access. In: Zhao, D., Du, J., Wang, H., Wang, P., Ji, D., Pan, J.Z. (eds.) CSWS 2014. CCIS, vol. 480, pp. 67–77. Springer, Heidelberg (2014)
4. Calvanese, D., De Giacomo, G., Lembo, D., Lenzerini, M., Rosati, R.: Tractable reasoning and efficient query answering in description logics: the DL-Lite family. J. Autom. Reason. **39**(3), 385–429 (2007)
5. Das, S., Sundara, S., Cyganiak, R.: R2RML: RDB to RDF Mapping Language. W3C RDB2RDF Working Group, W3C recommendation, September 2012
6. Giese, M., Soylu, A., Vega-Gorgojo, G., Waaler, A., Haase, P., Jiménez-Ruiz, E., Lanti, D., Rezk, M., Xiao, G., Özçep, Ö.L., Rosati, R.: Optique: zooming in on big data. IEEE Comput. **48**(3), 60–67 (2015)
7. Glimm, B., Horrocks, I., Motik, B., Stoilos, G., Wang, Z.: HermiT: an OWL 2 reasoner. J. Autom. Reason. **53**(3), 245–269 (2014)
8. Haase, P., Schmidt, M., Schwarte, A.: The information workbench as a self-service platform for linked data applications. In: Hartig, O., Harth, A., Sequeda, J. (eds.) Proceedings of the Second International Workshop on Consuming Linked Data (COLD2011), Bonn, Germany, October 23, 2011, vol. 782 of CEUR Workshop Proceedings. CEUR-WS.org (2011)
9. Kharlamov, E., et al.: Ontology based access to exploration data at Statoil. In: Arenas, M., et al. (eds.) ISWC 2015. LNCS, vol. 9367, pp. 93–112. Springer, Heidelberg (2015). doi:10.1007/978-3-319-25010-6_6
10. Lembo, D., Mora, J., Rosati, R., Savo, D.F., Thorstensen, E.: Towards mapping analysis in ontology-based data access. In: Kontchakov, R., Mugnier, M.-L. (eds.) RR 2014. LNCS, vol. 8741, pp. 108–123. Springer, Heidelberg (2014)
11. Lembo, D., Mora, J., Rosati, R., Savo, D.F., Thorstensen, E.: Mapping analysis in ontology-based data access: algorithms and complexity. In: Proceedings of the 14th International Semantic Web Conference (ISWC) (2015)
12. Poggi, A., Lembo, D., Calvanese, D., De Giacomo, G., Lenzerini, M., Rosati, R.: Linking data to ontologies. In: Spaccapietra, S. (ed.) Journal on Data Semantics X. LNCS, vol. 4900, pp. 133–173. Springer, Heidelberg (2008)

Validating Ontologies Against OWL 2 Profiles with the SPARQL Template Transformation Language

Olivier Corby[✉], Catherine Faron-Zucker[✉], and Raphaël Gazzotti

Université Côte d'Azur, Inria, CNRS, I3S, Nice, France
olivier.corby@inria.fr, faron@unice.fr, gazzotti@i3s.unice.fr

Abstract. In this paper we address the general research question of *How can we express constraints on RDF data and how can we check that an RDF graph satisfies some given constraints?* and we focus on expressing constraints defining OWL 2 profiles and checking these constraints for OWL validation. We propose an approach based on the SPARQL Template Transformation language (STTL). An STTL template is a transformation rule that applies to a given RDF graph and the recursive call of a set of STTL templates on an RDF graph outputs some textual data resulting from the transformation of this graph. We show that STTL can be used as a constraint language for RDF and we use it to implement the semantics of OWL 2 profiles: each profile is represented by a set of STTL templates that a valid ontology must satisfy.

1 Introduction

OWL 2 profiles [6] can be seen as restrictions of OWL 2 statements and the validation of ontologies against OWL 2 profiles as the checking of syntactic constraints on OWL 2 axiom declarations. In this paper we address the general research question of *How can we express constraints on RDF data and how can we check that an RDF graph satisfies some given constraints?* and we focus on expressing constraints defining OWL 2 profiles and checking these constraints for OWL validation.

We propose an approach based on the SPARQL Template Transformation language (STTL), which we originally designed in order to enable the transformation of RDF data into any data format. An STTL template can be viewed as a transformation rule that applies to a given RDF graph just like an XSL template applies to an XML tree, and the recursive call of a set of STTL templates on a whole RDF graph outputs some textual data resulting from the transformation of this graph.

We show that STTL can be used as a constraint language for RDF: each STTL template is viewed as representing a constraint and an RDF graph is checked against a set of constraints by applying the set of STTL templates representing these constraints on the RDF graph. The output of the application of a set of STTL templates can be a simple boolean value or a convenient textual

© Springer International Publishing Switzerland 2016
M. Ortiz and S. Schlobach (Eds.): RR 2016, LNCS 9898, pp. 39–45, 2016.
DOI: 10.1007/978-3-319-45276-0_4

view of the data, where for instance, the subgraphs violating the constraints are highlighted. This is done by defining a "Visitor" design patten associated to the set of STTL templates in order to collect illegal RDF sub-graphs, and a generic design pattern to display the result to the user.

As a result, we apply our approach to implement the semantics of OWL 2 profiles, each viewed as a set of constraints to be validated: we defined an STTL transformation to represent each of the three OWL 2 profiles (OWL RL, OWL QL and OWL EL). The application of one of these STTL transformations to an ontology (expressed in RDF) enables to validate it against the OWL 2 profile this transformation represents.

The paper is organized as follows. Section 2 provides an overview of the STTL language. Section 3 presents the STTL transformation implementing the semantics of the OWL 2 profiles. Section 4 shows how an additional STTL transformation enables to provide the user with a visual presentation of the results of the OWL validation results. Section 5 describes our experiments conducted on several OWL ontologies of the Linked Data. Section 6 concludes.

2   SPARQL Template Transformation Language (STTL)

STTL is a generic transformation rule language for RDF which relies on two extensions of SPARQL: an additional TEMPLATE query form to express transformation rules and extension functions to recursively call the processing of a template from another one. A TEMPLATE query is made of a standard WHERE clause and a TEMPLATE clause. The WHERE clause is the condition part of a rule, specifying the nodes in the RDF graph to be selected for the transformation. The TEMPLATE clause is the presentation part of the rule, specifying the output of the transformation performed on the solution sequence of the WHERE part. For instance, let us consider the OWL axiom stating that the class of parents is equivalent to the class of individuals having a person as child. Here are its expressions in Functional syntax and in Turtle:

```
EquivalentClasses(a:Parent
  ObjectSomeValuesFrom(a:hasChild a:Person))
```

```
a:Parent a owl:Class ; owl:equivalentClass
   [ a owl:Restriction ; owl:onProperty a:hasChild ;
     owl:someValuesFrom a:Person ]
```

The template below enables to transform the above `equivalentClass` statement from RDF into Functional syntax:

```
TEMPLATE { FORMAT {"EquivalentClasses(%s %s)"
     st:apply-templates(?in) st:apply-templates(?c) }}
WHERE { ?in owl:equivalentClass ?c }
```

The value matching variable ?in is a:Parent which is expected in the transformation output (the Functional syntax of the OWL 2 statement), while the value matching variable `?c` is a blank node whose property values are used to build the expected output. This is defined in another template to be applied on this focus node. The `st:apply-templates` extension function enables this recursive call of templates, where `st` is the prefix of STTL namespace[1].

More generally, function `st:apply-templates` can be used in the TEMPLATE clause of any template t_1 to execute another template t_2 that can itself execute a template t_3, etc. Hence, templates call themselves one another, in a series of calls, enabling a hierarchical processing of templates and a recursive traversing of the target RDF graph. The STTL interpreter keeps track of templates and focus nodes in order to prevent loops as RDF graphs may have cycles. Similarly, function `st:call-template` can be used to recursively call named templates.

STTL is compiled into standard SPARQL. The compilation keeps the WHERE clause, the solution modifiers and the VALUES clause of the template unchanged and the TEMPLATE clause is compiled into a SELECT clause.

A complete description of STTL language is provided in [1]. We implemented the STTL language and transformer engine within the Corese Semantic Web Factory [3] which now comprises an STTL RESTful Web service to process STTL transformations and output the result of transforming an RDF dataset. This implementation is described in [2].

3   Validating OWL 2 Profiles with STTL Transformations

OWL 2 profiles are logic fragments, or sublanguages, trading expressive representation power for efficient reasoning capabilities. There are three profiles predefined in the recommendation: EL, QL and RL. As stated in the W3C recommendation, each OWL 2 profile is defined as a set of restrictions on the structure of OWL 2 statements, i.e. syntactic constraints on OWL 2 axioms definitions[2]. Each profile is defined as (1) a set of restrictions on the type of class expressions that can be used in axioms and on the place in which they can be used, (2) the set of OWL axioms supported when restricted to the allowed set of class expressions, (3) the set of OWL constructs which are not supported. For example, in OWL 2 RL, the constructs in the subclass and superclass expressions in `SubClassOf` axioms must follow some usage patterns and OWL 2 RL axioms are undirectly constrained by these restrictions.

We defined an STTL transformation to represent each of the three OWL 2 profiles defined in the W3C recommendation. Each STTL template participating to these transformations enables to check a specific OWL 2 model constraint and returns a boolean, the value of which depends on whether the constraint is verified or not. When traversing the RDF graph representing the ontology to be validated against a given OWL 2 profile, the boolean results of the templates applied to the graph nodes are aggregated by using a conjunction instead of a concatenation, so that the final result is a boolean value indicating whether type checking succeeds or fails.

---

[1] http://ns.inria.fr/sparql-template/.
[2] https://www.w3.org/TR/owl2-profiles/.

Considering that each OWL 2 profile is defined by a set of constraints for the declaration of axioms (some axioms are not supported, some are supported with restrictions) and a set of constraints on class expressions, we defined *modular* STTL transformations to represent OWL 2 profiles. Basically, each one consists in a single template calling a transformation gathering templates representing constraints on axioms and these transformations call several other transformations gathering templates representing constraints on class expressions.

Let us focus on the `st:owlrl`[3] transformation which comprises 36 STTL templates representing the constraints defining the OWL 2 RL profile. It consists of a start template calling the `st:axiom` transformation whose templates themselves call the `st:subexp`, `st:superexp`, and `st:equivexp` transformations. Transformation `st:axiom` comprises 10 templates representing restrictions on class axioms to use the appropriate form of class expressions, restrictions on property domain and range axioms to only use class expressions of type `superClassExpression`, restriction on positive assertions to only use class expressions of type `superClassExpression` and restrictions on keys to only use `subClassExpression`.

The result of each template is a boolean value that represents the conformance of the axiom arguments. For instance, the following template represents the restriction on `subClassOf` axioms to use a class expression of type `superClassExpression` (respectively `subClassExpression`) for the superclass (respectively the subclass). These two types of class expressions are each defined by another STTL transformation which is recursively called in the WHERE clause of the template. More precisely, a `subClassOf` axiom is represented by an RDF triple whose property is `rdfs:subClassOf`, whose subject `?in` is passed as argument to transformation `st:subClassExpression` and whose object `?y` is passed as argument to transformation `st:superClassExpression`. Both transformations return a boolean whose value corresponds to the conformance of the class expressions. The template returns the conjunction of these two booleans. In addition, a "Visitor" design pattern is used to report axioms which are not conform to the profile.

```
TEMPLATE { ?suc }
WHERE {
  ?in rdfs:subClassOf ?y
  BIND (
    st:call-template-with(st:subexp, st:subClassExpression, ?in) &&
    st:call-template-with(st:superexp, st:superClassExpression, ?y)
  AS ?suc)
  FILTER st:alreadyVisited(?in,"subClass", ?suc) }
```

In addition, `st:axiom` comprises one template representing the disallowance of the `DisjointUnion` axiom and of reflexive properties. This template returns `false` if such an axiom or property occurs in the ontology.

---

[3] http://ns.inria.fr/sparql-template/owlrl/owlrl.

```
TEMPLATE { false }
WHERE {
  {?in owl:disjointUnionOf ?y} UNION {?in a owl:ReflexiveProperty}
  FILTER (st:alreadyVisited(?in,"fail", false)) } LIMIT 1
```

We defined an STTL transformation for each of the three types of class expressions in OWL 2 RL: `subClassExpression`, `superClassExpression` and `equivClassExpression`. For instance, let us consider the `st:subexp` transformation representing the `subClassExpression` type of class expressions, that can occur as subclass expressions in `SubClassOf` axioms. In this transformation, the following named template `st:subClassExpression` calls for all the other templates in the transformation. It enables to checks whether the argument is a URI, in which case it must not be `owl:Thing`; otherwise it checks whether all the templates matching the argument return true. In addition, a "Visitor" design pattern is used to report expressions that do not conform.

```
TEMPLATE st:subClassExpression(?x) { ?suc }
WHERE {
  BIND (
    IF (isURI(?x), ?x != owl:Thing, st:apply-templates-all(?x))
  AS ?suc)
  BIND (st:visit(st:sub, ?x, ?suc) as ?b) }
```

4   Validation Result Presentation

In order to provide the user with a visualization of the result of the validation, we wrote an STTL transformation to present in a HTML document the RDF graph (in the Turtle syntax) representing the ontology to be validated, where non valid triples are highlighted. For instance, Fig. 1 shows the visualization of an ontology represented in Turtle and tested against the OWL 2 RL profile with `owl:complementOf` in red since OWL 2 RL does not allow this within a class intersection inside a class equivalence.

During the traversal of the RDF graph representing the tested ontology, a visitor records the subjects of RDF triples corresponding to failing statements. After type check resumes, the visitor is given to an STTL transformation RDF2Turtle which enables to pretty-print RDF graphs in Turtle. The template below is the key of the STTL transformation. It uses the `st:visited(?in)` extension function which returns true if the node has been visited (and hence represents a failing statement). When processing a node of the RDF graph representing the vocabulary to be validated, in case this node represents a failing OWL statement, the STTL template generates a `<span class='fail'>` ... `</span>` HTML element to embed the transformation of the node, i.e. its pretty-print in Turtle embeded in HTML. A CSS stylesheet associates a specific presentation format to the `fail` class, e.g. a red font color.

```
<http://example.com/owl/families/ChildlessPerson>
  a owl:Class ;
  rdfs:subClassOf [a owl:Class ;
    owl:intersectionOf (<http://example.com/owl/families/Person>
    [a owl:Class ;
      owl:complementOf [a owl:Restriction ;
        owl:onProperty [a owl:ObjectProperty ;
          owl:inverseOf <http://example.com/owl/families/hasParent>] ;
        owl:someValuesFrom owl:Thing]])] ;
  owl:equivalentClass [a owl:Class ;
    owl:intersectionOf ([a owl:Class ;
      owl:complementOf <http://example.com/owl/families/Parent>]
      "OWL RL: Statement not supported in an Equivalent Class Expression."

    <http://example.com/owl/families/Person>)]
    "OWL RL: Class Expression not supported with owl:equivalentClass or owl:intersectionOf."
```

Fig. 1. Visualizing the validation result of an ontology against OWL 2 RL

```
TEMPLATE { FORMAT {
  if (st:visited(?in),"[<span class='fail'>%s</span>].","[%s].")
  ibox { st:call-template(st:type, ?in)
         st:call-template(st:value, ?in) } }}
WHERE { ?in ?p ?y  FILTER isBlank(?in) } LIMIT 1
```

5   Implementation and Experiments

We have written an STTL transformation for the three OWL profiles defined in the W3C recommendation: OWL RL (36 templates), OWL QL (24 templates) and OWL EL (20 templates)[4]. These transformations, like any other STTL transformations, can be applied to an OWL ontology to be validated by using the Corese Semantic Web Factory which comprises an STTL engine. This is an open-source development that can be freely downloaded[5]. We also wrote and deployed a dedicated Web service that can validate an OWL ontology against OWL 2 profiles given the URL of the ontology as an argument in the HTTP request (in RDF)[6]. We also have tested the STTL transformations on a proprietary ontology in the e-Education domain, owned by the Educlever company. It comprises 57,174 triples and its validation took 0.5 s on a laptop (HP EliteBook 840 G2, 2.6 GHz, 16 GB RAM). Finally, we have tested the STTL constraint checking transformations on the open source Foundational Model of Anatomy (FMA) ontology[7]. It comprises 1,743,162 triples and its validation against OWL RL takes 3.3 s, against OWL QL 4.8 s, and against OWL EL 4.6 s.

6   Conclusion

We have shown how to answer the problem of OWL 2 RL Profile conformance checking by using the STTL language. We have designed an STTL

---

[4] http://ns.inria.fr/sparql-template/.

[5] http://wimmics.inria.fr/corese.

[6] http://corese.inria.fr/.

[7] http://sig.biostr.washington.edu/projects/fma/release/index.html.

transformation for each of the OWL 2 profiles in the W3C recommendation. The STTL engine as well as the STTL transformations are freely available and open-source and a Web service enables to test our validators with any ontology (in RDF). We have created a design pattern that enables transformations to perform type checking by returning boolean values and pretty-print the result of the validation. As future work, we will provide a comparison of our OWL 2 validator to the validator developed by the University of Manchester[8] which relies on the OWL API [5].

Our approach to represent OWL 2 profiles by STTL transformations is not specific to the problem of OWL validation and STTL can be used to represent other kinds of constraints on RDF data. Therefore, as future work, we will compare our approach to related works on RDF constraint checking, among which [4]. Relatedly, the W3C hosts a RDF Data Shapes[9] working group for describing structural constraints and validate RDF data against those and we are currently designing an STTL transformation implementing the current version of W3C RDF Data Shapes.

References

1. Corby, O., Faron-Zucker, C.: STTL: a SPARQL-based transformation language for RDF. In: 11th International Conference on Web Information Systems and Technologies, WEBIST 2015, Lisbon, Portugal, May 2015
2. Corby, O., Faron-Zucker, C., Gandon, F.: A generic RDF transformation software and its application to an online translation service for common languages of linked data. In: Arenas, M., et al. (eds.) ISWC 2015. LNCS, vol. 9367, pp. 150–165. Springer, Heidelberg (2015)
3. Corby, O., Gaignard, A., Faron-Zucker, C., Montagnat, J.: KGRAM versatile data graphs querying and inference engine. In: IEEE/WIC/ACM International Conference on Web Intelligence, Macau, China (2012)
4. Fischer, P.M., Lausen, G., Schätzle, A., Schmidt, M.: RDF constraint checking. In: Proceedings of the Workshops of the EDBT/ICDT 2015 Joint Conference. CEUR Workshop Proceedings, Brussels, Belgium, vol. 1330, pp. 205–212 (2015)
5. Horridge, M., Bechhofer, S.: The OWL API: a Java API for OWL ontologies. Semantic Web J. **2**, 11–21 (2011)
6. Motik, B., Grau, B.C., Horrocks, I., Zhe, W., Fokoue, A., Lutz, C.: OWL 2 Web ontology language profiles. Recommendation, W3C (2012). http://www.w3.org/TR/owl2-profiles/

---

[8] http://mowl-power.cs.man.ac.uk:8080/validator/.
[9] http://www.w3.org/2014/data-shapes/wiki/Main_Page.

Revisiting Grounded Circumscription in Description Logics

Stathis Delivorias[2(✉)] and Sebastian Rudolph[1]

[1] Theoretical Computer Science, TU Dresden, Dresden, Germany
Sebastian.Rudolph@tu-dresden.de
[2] University of Montpellier, LIRMM, Montpellier, France
Delivorias@lirmm.fr

Abstract. Circumscription is a paradigm of non-monotonic logic meant to formalize the common-sense understanding that, among competing theories that represent phenomena equally well, the one with the fewest "abnormal" assumptions should be selected. Several papers have considered ways of adding circumscription to Description Logics. One of the proposals with good computational properties is Grounded Circumscription, introduced by Sengupta, Krishnadi and Hitzler in 2011. Our paper builds on their general idea, but identifies some problems with the original semantics definition, which gives rise to counter-intuitive consequences and renders the proposed tableau algorithm incorrect. We give an example that makes the problem explicit and propose a modification of the semantics that remedies this issue. On the algorithmic side, we show that a big part of the reasoning can actually be transferred to standard Description Logics, for which tools and results already exist.

1 Introduction

Circumscription is a paradigm of non-monotonic logic introduced by John McCarthy in 1980 [5]. The main idea is to formalize the common sense understanding that among competing theories that predict equally well, the one with the fewest assumptions should be selected. This is basically an application of the principle known as "Occam's razor" to logic. It is also similar to the closed world assumption, where what is not known to be true is taken to be false. In its original first-order logic formulation, circumscription minimizes the extension of some predicates, where the extension of a predicate is the set of tuples of values the predicate is true on.

Description Logics (DLs) are knowledge representation formalisms designed to describe and reason about qualitative properties and conceptual aspects of a system [1,6]. Ontology languages based on DLs have been widely adopted in a large class of application areas. One of the most prominent applications of DLs is to provide the underlying logical basis of the web ontology language OWL 2, which is the current recommendation of the World Wide Web Consortium (W3C) [4,8]. Therein DLs are used to represent the intended meaning of Web resources and establish powerful reasoning tools, so as to facilitate machine

© Springer International Publishing Switzerland 2016
M. Ortiz and S. Schlobach (Eds.): RR 2016, LNCS 9898, pp. 46–60, 2016.
DOI: 10.1007/978-3-319-45276-0_5

understandability of Web pages. From a more scholarly perspective, DLs are decidable fragments of first order logic.

Description Logics traditionally operate within the monotonic realm, namely the addition of more assertions to a knowledge base does not negate previously inferred information. But in many prevalent application domains, such as common sense reasoning, this property does not hold. Conclusions might need to be revised in the light of new information. Hence it is quite intriguing to try to develop a DL framework where reasoning would be non-monotonic. There have been notable efforts to define circumscription for DLs, albeit with rather high complexity or even undecidable if roles are circumscribed [2].

In this essay we aim to fuse Description Logics, with a restricted version of Circumscription, called Grounded Circumscription. The work is based on a 2011 publication by K. Sengupta, A.A. Krisnadhi and P. Hitzler, which throughout this work we will refer to as "the original paper" [7]. In ground circumscription, some of the predicates in our language (which in DL can only be unary or binary) are chosen to be grounded and minimized. Grounded means that their interpretations must include only named individuals, i.e. elements of the domain that correspond to one of the constants that appear in our knowledge base. Moreover, those predicates are minimized in the sense that we accept only models which assign as few individuals as possible to them, so that there cannot be a model whose extensions of these predicates are subsets of the respective extensions in the minimal model.

In the original paper, the main idea of grounded circumscription is given along with algorithms for certain decision problems. We have optimized and modified these ideas. The optimization was our initial aim, in particular we wanted (and largely achieved) to transfer a big part of the reasoning to standard DLs, for which there already exist tools and available results. But in the process we uncovered some insufficiencies in the notion of minimality as introduced in the original paper, to the discussion of which Subsect. 3.1 is devoted. Hence we have modified the main definition to one that is more effective and more intuitive.

After introducing the particular DL formalism and terminology that we work on (Sect. 2), we specify the basic notions and proceed to present an algorithm for satisfiability (Sect. 3) which is predominantly in the monotonic sphere. We then introduce important notions which are put to use in the algorithm for entailment of facts (Sect. 4). Following are supplementary results that further develop the theory of grounded circumscription in DLs (Sect. 5) and finally we give an overview of the contribution of this endeavor and discuss prospects of further research (Sect. 6).

All proofs can be found in the original master thesis [3].

2 Preliminaries

In this section, we give a brief introduction to our formalism and the main terminology and ideas around it.

In the original paper, decidability of ground circumscription is proven using rather complex and non-standard languages which feature concept products, role hierarchies and role disjunctions. Then, independently, algorithms which apply only to $\mathcal{ALC}$ are given. Our work is entirely based on the standard DL $\mathcal{ALCO}$ but it can trivially be extended to any more complex formalism that subsumes $\mathcal{ALCO}$.

$\mathcal{ALCO}$ **Syntax.** Let N_C, N_r and N_I be mutually disjoint sets of *concept-*, *role-* and *individual names*, respectively. Concepts C in $\mathcal{ALCO}$ are built using the grammar rule:

$$C ::= \top \mid A \mid \{a\} \mid \neg C \mid C \sqcap C \mid \exists r.C$$

where $A \in N_C$, $r \in N_r$, and $a \in N_I$. We employ the usual abbreviations: $\bot = \neg\top$, $C \sqcup D = \neg(\neg C \sqcap \neg D)$, and $\forall r.C = \neg\exists r.\neg C$.

An expression of the form $C \sqsubseteq D$, where C and D are concepts, is called a *general concept inclusion* (GCI). A finite set of GCIs is a *TBox*. An expression of the form $C(a)$, where C is a concept and $a \in N_I$, is called a *concept assertion*. For $r \in N_R$ and $a, b \in N_I$, an expression of the form $r(a,b)$ is called a *role assertion*. A finite set of concept and role assertions is called an *ABox*.

A pair $K = (\mathcal{T}, \mathcal{A})$ consisting of a TBox $\mathcal{T}$ and an ABox $\mathcal{A}$ is called a *knowledge base* (abbreviated frequently as KB). For ease of presentation, in this study we will usually understand a knowledge base as a single set of axioms, which would formally be expressed as $K = \mathcal{T} \cup \mathcal{A}$. We will not refer to Aboxes and Tboxes individually, rather we will handle the knowledge base as a whole.

$\mathcal{ALCO}$ **Semantics.** An *interpretation* is a pair $\mathcal{I} = (\Delta, \cdot^{\mathcal{I}})$, where Δ is a non-empty *domain* and $\cdot^{\mathcal{I}}$ is a function that maps every $a \in N_I$ to $a^{\mathcal{I}} \in \Delta$, every $A \in N_C$ to $A^{\mathcal{I}} \subseteq \Delta$, and every $r \in N_r$ to $r^{\mathcal{I}} \subseteq \Delta \times \Delta$. The mapping $\cdot^{\mathcal{I}}$ is naturally extended to all concepts by setting

$$\top^{\mathcal{I}} = \Delta,$$
$$(\neg C)^{\mathcal{I}} = \Delta \setminus C^{\mathcal{I}},$$
$$(C \sqcap D)^{\mathcal{I}} = C^{\mathcal{I}} \cap D^{\mathcal{I}},$$
$$\{a\}^{\mathcal{I}} = \{a^{\mathcal{I}}\},$$
$$(\exists r.C)^{\mathcal{I}} = \{x \in \Delta \mid \exists y \in \Delta. (x,y) \in r^{\mathcal{I}} \wedge y \in C^{\mathcal{I}}\}.$$

An interpretation $\mathcal{I}$ *satisfies*

- a concept inclusion $C \sqsubseteq D$ if $C^{\mathcal{I}} \subseteq D^{\mathcal{I}}$,
- a concept assertion $C(a)$ if $a^{\mathcal{I}} \in C^{\mathcal{I}}$ and
- a role assertion $r(a,b)$ if $(a^{\mathcal{I}}, b^{\mathcal{I}}) \in r^{\mathcal{I}}$.

We say that $\mathcal{I}$ is a *model* of a TBox $\mathcal{T}$ or an ABox $\mathcal{A}$ if it satisfies every concept inclusion in $\mathcal{T}$ or every assertion in $\mathcal{A}$, respectively. $\mathcal{I}$ is a model of a knowledge base $K = (\mathcal{T}, \mathcal{A})$ if $\mathcal{I}$ is a model of both $\mathcal{T}$ and $\mathcal{A}$. If there exists a model of a knowledge base K, then K is a *satisfiable* KB. If every model of K satisfies $C(a)$ (or $r(a,b)$, respectively), we say that $C(a)$ (or $r(a,b)$, respectively) is *entailed* by K. If every model of K satisfies $C \sqsubseteq D$, then C is *subsumed* by D with respect to K.

3   Grounded Circumscription

In this section we formally define the basic notions of ground circumscription. The definition of minimality is reestablished in solid grounds, which can prove a useful framework for further development of this theory.

Ground Extension.   A central notion in this study is that of ground extension of a predicate with respect to a certain interpretation, which is the set of individual names or pairs of individual names (depending on whether the predicate is a concept or a role), whose interpretations belong to the interpretation of this predicate. Given a knowledge base K, the set of individual names that appear in K are symbolized $Ind(K)$.

Definition 1. Let K be an $\mathcal{ALCO}$ knowledge base and $\mathcal{I}$ an interpretation. The *ground extension* wrt $\mathcal{I}$ of a predicate $W \in N_C \cup N_r$, is the following set:

$$Ext^{\mathcal{I}}(W) := \begin{cases} \{a \in Ind(K) | a^{\mathcal{I}} \in W^{\mathcal{I}}\} & if\ W \in N_C, \\ \{(a,b) \in Ind(K) \times Ind(K) | (a^{\mathcal{I}}, b^{\mathcal{I}}) \in W^{\mathcal{I}}\} & if\ W \in N_r. \end{cases} \dashv$$

The key role that ground extension plays, is evident by its frequent presence throughout the rest of this work. $Ext^{\mathcal{I}}(\cdot)$ can be naturally extended to be applicable to any concept description: if C is a concept, then $Ext^{\mathcal{I}}(C) := \{a \in Ind(K) | a^{\mathcal{I}} \in C^{\mathcal{I}}\}$. Since nominals are valid concept constructors in our language, we can ultimately view $Ext^{\mathcal{I}}(C)$ as a concept description, provided that C is a concept as well. One more important property of ground extension, which is easy to verify, is that it is monotonic with respect to set inclusion, i.e. if $A^{\mathcal{I}} \subseteq B^{\mathcal{I}}$ then $Ext^{\mathcal{I}}(A) \subseteq Ext^{\mathcal{I}}(B)$.

Groundedness and Minimality.   The main idea in grounded circumscription is to select some predicates (concept and role names), and demand that for every model their interpretation is grounded, i.e. it includes only named individuals, and that it is minimized, in the sense that there cannot be an interpretation that assigns fewer individuals to those predicates and still is a model of our given knowledge base.

Definition 2. Let K be a knowledge base and $M \subseteq N_C \cup N_r$. A model $\mathcal{I}$ of K is called *grounded* wrt M if

(i) $C^{\mathcal{I}} \subseteq \{b^{\mathcal{I}} | b \in Ind(K)\}$ for every $C \in M \cap N_C$.
(ii) $r^{\mathcal{I}} \subseteq \{(a^{\mathcal{I}}, b^{\mathcal{I}}) | a, b \in Ind(K)\}$ for every $r \in M \cap N_r$. $\dashv$

Definition 3. A *GC-$\mathcal{ALCO}$-KB* is a pair (K, M) where K is an $\mathcal{ALCO}$ knowledge base and $M \subseteq N_C \cup N_r$. Every $W \in M$ is said to be *closed* wrt K. Let $\prec_M$ denote a *"smaller than"* relation which is a partial order on the set of all interpretations for K. An interpretation $\mathcal{I}$ is a *GC-model* of (K, M) if it is a grounded model of K wrt M and $\mathcal{I}$ is minimal wrt $\prec_M$, i.e. there is no grounded model $\mathcal{J}$ of K such that $\mathcal{J} \prec_M \mathcal{I}$. (K, M) is *satisfiable* if it has a GC-model. A statement ϕ is a *logical consequence* of (K, M) if every GC-model of (K, M) satisfies ϕ. We then say that (K, M) *entails* ϕ. $\dashv$

Note that ϕ in the above definition could be a GCI, a concept assertion or a role assertion. Obviously, the precise semantics depends on the concrete choice of the "smaller than" relation $\prec_M$, which will be discussed in the next paragraph. Henceforth we will frequently substitute the term GC-model, with *minimal grounded model* or simply *minimal model*.

3.1   Discussion on the Modification of the GC-Definition

The original paper employed the following definition for the "smaller-than" relation.

Definition 4. Let (K, M) be a *GC-$\mathcal{ALCO}$-KB*. If $\mathcal{I}$ and $\mathcal{J}$ are interpretations of K then the *"smaller than"* relation is defined in the following way:
$$\mathcal{I} \prec_M^{\mathrm{orig}} \mathcal{J} \text{ if}$$

(i) $\Delta^{\mathcal{I}} = \Delta^{\mathcal{J}}$ and $a^{\mathcal{I}} = a^{\mathcal{J}}$ for every $a \in Ind(K)$,
(ii) $W^{\mathcal{I}} \subseteq W^{\mathcal{J}}$ for every $W \in M$ and
(iii) there is a $W \in M$ such that $W^{\mathcal{I}} \subset W^{\mathcal{J}}$. ⊣

The first condition for the minimality relation in the original paper requires that two interpretations have equal domains in order for them to be comparable. Firstly, we argue that this is counter-intuitive. When we say that a model has fewer assumptions than another model, this does not imply any similarity of their domains, it rather requires that those predicates which are of importance to us are of smaller extension. Furthermore, the original definition imposes algorithmic problems as one will have to look for a minimal model for every possible domain cardinality. This can make devising correct procedures for expressive languages significantly more difficult.

In particular, in the original paper the proposed algorithm for instance checking in $\mathcal{ALC}$ does not take the definition fully into account. We present here an example that demonstrates the potential counter-intuitive results of the above definition, as well as its disagreement with the proposed algorithm.

Consider the following knowledge base:

GoodPerson ⊓ Murderer ⊑ Abnormal
          (a good person that is a murderer is an abnormal person)
⊤ ⊑ ∃r_1.(¬GoodPerson ⊓ ¬Murderer)
         (there exists someone who is not a good person nor a murderer)
⊤ ⊑ ∃r_2.GoodPerson   *(there exists a good person)*
⊤ ⊑ ∃r_3.Murderer   *(there exists a murderer)*
GoodPerson(Sam)   *(Sam is a good person)*

Assume the set of closed predicates is $M = \{\texttt{Abnormal}\}$. Then an expected consequence under the grounded circumscription semantics would be ¬Murderer(Sam).

Yet, we observe that the following interpretation $\mathcal{I}$ is a GC-model according to the original definition (and hence prevents the expected consequence):

$$\Delta^{\mathcal{I}} = \{1, 2\}$$
$$\texttt{Sam}^{\mathcal{I}} = 1$$
$$\texttt{GoodPerson}^{\mathcal{I}} = \texttt{Murderer}^{\mathcal{I}} = \texttt{Abnormal}^{\mathcal{I}} = \{1\}$$
$$\texttt{r}_1{}^{\mathcal{I}} = \{(1, 2), (2, 2)\}$$
$$\texttt{r}_2{}^{\mathcal{I}} = \texttt{r}_3{}^{\mathcal{I}} = \{(1, 1), (2, 1)\}$$

The only reason why $\mathcal{I}$ is a GC-model is because it is minimal among all models of cardinality 2. For models of greater domain sizes, $\texttt{Sam}$ would never be included in $\texttt{Murderer}$, since he is in $\texttt{GoodPerson}$ and we want to minimize $\texttt{Abnormal}$. Moreover, we note that this model is not produced by the *GC-model-finder* algorithm, given in the original paper, hence contradicting their definition. We believe that this is because the notion of GC-model was not meant to include an interpretation like $\mathcal{I}$.

We propose to overcome the problems with the original definition of the grounded circumscription semantics by modifying the notion of minimality in the following way.

Definition 5. If $\mathcal{I}$ and $\mathcal{J}$ are models of K then the *"smaller than"* relation is defined in the following way: $\mathcal{I} \prec_M^{new} \mathcal{J}$ if

(i) $Ext^{\mathcal{I}}(\{a\}) = Ext^{\mathcal{J}}(\{a\})$ for every $a \in Ind(K)$,
(ii) $Ext^{\mathcal{I}}(W) \subseteq Ext^{\mathcal{J}}(W)$ for every $W \in M$ and
(iii) there is a $W \in M$ such that $Ext^{\mathcal{I}}(W) \subset Ext^{\mathcal{J}}(W)$. ⊣

Our definition of minimality, which subsumes the one in the original paper (i.e. $\mathcal{I} \prec_M^{orig} \mathcal{J}$ implies $\mathcal{I} \prec_M^{new} \mathcal{J}$), is more intuitive in that it directly involves the assignment of individuals to concepts and roles. This is anyway at the heart of the tableau method used by the authors of the original paper when providing algorithms for the reasoning tasks in ground circumscription. Apart from being more intuitive, this is also the more realistic approach. Comparison between two models can simply be done on the basis of the mapping of individuals to concepts and roles. Although we acknowledge that requiring same domains in order to allow comparison between two models has been the default approach to circumscription in DLs so far, it seems to be an unnecessary specialization.

Furthermore, we will present in the following how we can significantly improve the inferencing algorithms in comparison to the original paper. Keeping the old definition would have hindered this development. Hence we believe that our reformulation of the notion of minimality is an improvement compared to the previous work. It is more efficient in producing results by avoiding to interfere as much with the actual semantics, whilst capturing the essense of the idea of ground circumscription in more satisfactory way.

3.2 Satisfiability of a GC-Knowledge Base

We present now a direct and complexitywise cheap way of determining whether a GC-$\mathcal{ALCO}$ knowledge base is satisfiable. To this end, we first enhance the

KB with axioms that ensure grounding of the closed predicates and then we take advantage of the following lemma, which effectively says that if a grounded model exists, then a minimal grounded model must exist as well.

Lemma 1. Both relations $\prec_M^{\text{orig}}$ and $\prec_M^{\text{new}}$ are well-founded on the class of grounded models of a knowledge base K wrt to M.   ⊣

Definition 6. Let (K, M) be a $GC\text{-}\mathcal{ALCO}\text{-}KB$, where $M \cap N_C = \{A_1, ..., A_n\}$ and $M \cap N_r = \{r_1, ..., r_m\}$. We define K_M as the $\mathcal{ALCO}\text{-}KB$ which consists of all the axioms that are included in K as well as the following ones:

- $P \equiv \{x | x \in Ind(K)\}$ where P is a fresh concept name,
- $A_i \sqsubseteq P$ for every $i \in \{1, ..., n\}$,
- $\exists r_j.\top \sqsubseteq P$ for every $j \in \{1, ..., m\}$,
- $\top \sqsubseteq \forall r_j.P$ for every $j \in \{1, ..., m\}$.

K_M is then called a *grounded* $\mathcal{ALCO}$ knowledge base.   ⊣

We do not give an explicit algorithm for determining satisfiability of a GC knowledge base, because we will show that this decision problem can be reduced to the satisfiability checking of a (standard) $\mathcal{ALCO}$ knowledge base. To solve the reasoning tasks of ground circumscription with the use of the already developed monotonic DL reasoning tools was our aim, and as the next proposition shows, in this case it is proven to be achieved quite ideally.

Proposition 1. Let (K, M) be a $GC\text{-}\mathcal{ALCO}\text{-}KB$. (K, M) is satisfiable (under the grounded circumscription semantics, both w.r.t. $\prec_M^{\text{orig}}$ and $\prec_M^{\text{new}}$) if and only if the $\mathcal{ALCO}$ knowledge base K_M is (classically) satisfiable.   ⊣

One observation worth mentioning here is that grounding, although defined at a semantic level, can be internalized in the syntax and expressed as a particular class of knowledge bases. And through this grounding, a localization of the non-monotonicity is achieved, such that for the principal task of deciding satisfiability, we do not even need to expand reasoning beyond the already known algorithms that exist for standard DLs.

4   Instance Checking

For the task of determining whether or not a concept assertion (also referred to as 'fact') is entailed by a GC-$\mathcal{ALCO}$ knowledge base, knowing that a minimal model exists is not enough. We have to be able to find this model or at least to negate the possibility of a grounded model being minimal. What seems to be more efficient is a bottom-up approach, where the grounded models found first are definitely minimal. From now on we are working only on our definition of minimality and we will write $\prec_M$ instead of $\prec_M^{\text{new}}$. Also in the following $Part(X)$ will denote the set of all partitions of a set X and $\mathbb{Z}_2^* = \{0, 1\}^*$ will denote the set of all finite binary words.

Specification of the Configuration Space. The idea of defining independently what is essentially the search space of our algorithm, a space of possible choices of extensions to the closed predicates, is inspired by the original paper, where a similar set is specified. However, and this is one more clue which points to the divergence between the intended meaning of ground circumscription and what was initially defined, in the original paper the domain and the possible interpretations of the individuals over it are not taken into consideration when defining this space. Having improved the definition, we still need to add a dimension to the search space which will correspond to the possible interpretations of the individual names.

Given an interpretation $\mathcal{I}$, the *individual allocation* of $\mathcal{I}$ is the set $\mathsf{AL}(\mathcal{I}) \in Part(Ind(K))$ such that every $X \in \mathsf{AL}(\mathcal{I})$ has the property: for every $a \in X$ and $b \in Ind(K)$ holds that $a^{\mathcal{I}} = b^{\mathcal{I}}$ if and only if $b \in X$.

Suppose that $I \in Part(Ind(K))$ and $a, b \in Ind(K)$. We call a and b *I-invariant* and write $a \simeq_I b$ if there is an $X \in I$ such that $a, b \in X$. A set $Z \subseteq Ind(K)$ is called *I-complete* if $a \in Z$ and $a \simeq_I b$ imply $b \in Z$. Similarly a set $V \in Ind^2(K)$ is called *I-complete* if $(a, b) \in V$ and $a \simeq_I a'$ and $b \simeq_I b'$ imply $(a', b') \in V$. For the sake of conciseness in the next definition, we define the following sets:

$$Cmp_I(K) = \{X \subseteq Ind(K) | X \text{ is } I\text{-complete}\}$$
$$Cmp_I^2(K) = \{Y \subseteq Ind^2(K) | Y \text{ is } I\text{-complete}\}$$

We can now employ the above notions to specify the search space of our algorithm:

Definition 7. Let (K, M) be a $GC\text{-}\mathcal{ALCO}\text{-}KB$, where $M \cap N_C = \{A_1, ..., A_n\}$ and $M \cap N_r = \{r_1, ..., r_m\}$. Then the set

$$\mathcal{G}_{(K,M)} = \Big\{(X_1, ..., X_n, Y_1, ..., Y_m, I) \Big| X_i \subseteq Cmp_I(K), Y_j \subseteq Cmp_I^2(K), I \in$$
$$Part(Ind(K))\Big\}$$

is called *configuration space* of (K, M). ⊣

$\mathcal{G}_{(K,M)}$ is obviously a finite set. Every grounded model $\mathcal{I}$ of K wrt. to M corresponds to a point in the configuration space. In particular we will call the tuple

$$\Big(Ext^{\mathcal{I}}(A_1), ..., Ext^{\mathcal{I}}(A_n), Ext^{\mathcal{I}}(r_1), ..., Ext^{\mathcal{I}}(r_m), \mathsf{AL}(\mathcal{I})\Big)$$

the *assignment* of $\mathcal{I}$.

Let $G_1, G_2 \in \mathcal{G}_{(K,M)}$ with $G_1 = (Z_1, ..., Z_{n+m}, I)$ and $G_2 = (V_1, ..., V_{n+m}, I)$. We say that G_1 is *smaller than* G_2 and we write $G_1 \prec G_2$ if it holds that $Z_i \subseteq V_i$ for all $i \in \{1, ..., n + m\}$ and there exists $i \in \{1, ..., n + m\}$ such that $Z_i \subset V_i$. The following result then holds trivially:

Lemma 2. Let $\mathcal{I}, \mathcal{J}$ be grounded models of a knowledge base K wrt M and let $G_1, G_2 \in \mathcal{G}_{(K,M)}$ be their respective assignments. Then it holds that $\mathcal{I} \prec_M \mathcal{J}$ if and only if $G_1 \prec G_2$. ⊣

Binary Encoding and Linear Order. Let $\mathcal{G}_{(K,M)}$ be the configuration space of a $GC\text{-}\mathcal{ALCO}\text{-}KB$. Let $Ind(K) = \{a_1, ..., a_\mu\}$. For the purposes of the algorithm presented in the next section, we want to order $\mathcal{G}_{(K,M)}$ linearly. We achieve that by using a binary encoding for every $G \in \mathcal{G}_{(K,M)}$ and the lexicographical order. We first introduce an encoding $s : Part(Ind(K)) \to \mathbb{Z}_2^*$. Every partition of $Ind(K)$ can be specified by indicating which couples of individual names (that appear in the knowledge base) belong to the same block of the partition. This is easily percieved with the following visualization:

	a_1	a_2	a_3	$\cdots$	a_μ
a_1	-	$s_{(1,2)}$	$s_{(1,3)}$	$\cdots$	$s_{(1,\mu)}$
a_2	-	-	$s_{(2,3)}$	$\cdots$	$s_{(2,\mu)}$
$\vdots$	-	-	-	$\ddots$	$\vdots$
$a_{\mu-1}$	-	-	-	-	$s_{(\mu-1,\mu)}$
a_μ	-	-	-	-	-

In accordance with the above table we define

$$s(I) = s_{(1,2)}s_{(1,3)}\cdots s_{(1,\mu)}s_{(2,3)}s_{(2,4)}\cdots s_{(2,\mu)}\cdots s_{(\mu-1,\mu)}$$

where $s_{(i,j)} = 1$ if there exists $Z \in I$ with $a_i, a_j \in Z$, otherwise $s_{(i,j)} = 0$. We can now proceed to define the complete binary encoding of the points of the configuration space.

Let $\sigma : \mathscr{P}\Big(Ind(K)\Big) \cup \mathscr{P}\Big(Ind^2(K)\Big) \cup Part\Big(Ind(K)\Big) \to \mathbb{Z}_2^*$, with

$$\sigma(X) = \begin{cases} z_1 z_2 \ldots z_\mu & \text{if } X \subseteq Ind(K) \text{ where } z_\kappa = \begin{cases} 1 & \text{if } a_\kappa \in X \\ 0 & \text{if } a_\kappa \notin X \end{cases} \\ z_1 z_2 \ldots z_{\mu^2} & \text{if } X \subseteq Ind^2(K) \text{ where } z_{\kappa\mu+\lambda} = \begin{cases} 1 & \text{if } (a_\kappa, a_\lambda) \in X \\ 0 & \text{if } (a_\kappa, a_\lambda) \notin X \end{cases} \\ s(X) & \text{if } X \in Part(Ind(K)) \end{cases}$$

We can view words over $\mathbb{Z}_2$ as natural numbers encoded in the binary system. If $w_1, w_2 \in \mathbb{Z}_2^*$ are words of the same length, we write $w_1 < w_2$ if this relation holds for the respective natural numbers. We can now define a total order '$<$' on $\mathcal{G}_{(K,M)}$.

Definition 8. Let $G_1 = (Z_1, ..., Z_k)$ and $G_2 = (V_1, ..., V_k)$ be two points in the configuration space of a $GC\text{-}\mathcal{ALCO}\text{-}KB$ (K, M). G_1 *precedes* G_2 and we write $G_1 < G_2$ if there exists an $i \leq k$ such that $\sigma(Z_i) < \sigma(V_i)$ and for all $j < i$ holds $\sigma(Z_j) = \sigma(V_j)$. $\dashv$

For efficiency purposes, it is important here that the order defined above is induced by the partial order of minimality, so that the algorithm will discover the minimal model first and discard searching in large sections of the configuration space. The following lemma ensures that this is indeed the case.

Lemma 3. Let $\mathcal{G}_{(K,M)}$ be the configuration space of a $GC\text{-}\mathcal{ALCO}\text{-}KB$. For every $G_1, G_2 \in \mathcal{G}_{(K,M)}$ holds that $G_1 \prec G_2$ implies $G_1 < G_2$.   ⊣

Navigation Within the Configuration Space.   It is critical, given a point in the configuration space, to be able to construct a grounded model with such an assignment, if one exists. This is accomplished by adding the axioms specified in the next definition.

Definition 9. Let (K, M) be a $GC\text{-}\mathcal{ALCO}\text{-}KB$, where $M \cap N_C = \{A_1, ..., A_n\}$ and $M \cap N_r = \{r_1, ..., r_m\}$. Let $G = (X_1, ..., X_n, Y_1, ..., Y_m, I)$ be a point in the configuration space of (K, M). We define K_G as the $\mathcal{ALCO}-KB$ which consists of all the axioms that are included in K_M as well as the following ones:

- $\{a\} \equiv \{b\}$ for all $a, b \in Ind(K)$ with $a \simeq_I b$,
- $\{a\} \sqsubseteq \neg\{b\}$ for all $a, b \in Ind(K)$ with $a \not\simeq_I b$,
- $A_i \equiv X_i$ for every $i \in \{1, ..., n\}$,
- $N_{(a,j)} \equiv \{c \in Ind(K) | (c, a) \in Y_j\}$ for every $a \in Ind(K)$ and $j \in \{1, ..., m\}$,
- $\exists r_j.\{a\} \equiv N_{(a,j)}$ for every $a \in Ind(K)$ and $j \in \{1, ..., m\}$.

K_G is then called a *pointwise restriction* of (K, M).   ⊣

Lemma 4. Let K_G be a pointwise restriction of a $GC\text{-}\mathcal{ALCO}$ knowledge base (K, M). The following statements hold:

(i) If $\mathcal{I}$ is a model of K_G, then G is the assignment of $\mathcal{I}$.
(ii) If there exists a model of K_M with assignment G, then K_G is satisfiable.   ⊣

The Algorithm.   We can now specify an algorithm for deciding whether or not a GC knowledge base entails an assertion. We will also give an example and discuss complexity and possibility of further use and development.

Let (K, M) be a $GC\text{-}\mathcal{ALCO}\text{-}KB$, where $M \cap N_C = \{A_1, ..., A_n\}$ and $M \cap N_r = \{r_1, ..., r_m\}$. We want to check if an assertion $B(a)$ is a logical consequence of (K, M). Such a reasoning task is commonly refered to as *instance checking*, hence the title of this section. If $a \notin Ind(K)$ the answer is trivial, so for the rest it assumed that $a \in Ind(K)$. We split the decision procedure in two cases, the first of which will prove to be solvable in a much more simple way, by only once calling the "oracle" $\mathcal{ALCO}$ reasoner. In the following, given a knowledge base K_0, we use the notation $K_0^+ := K_0 \cup \{\neg B(a)\}$ to refer to K_0 augmented with the negation of the assertion we are checking for entailment.

Case 1: $B \in M$

Proposition 2. K_M^+ is unsatisfiable if and only if (K, M) entails $B(a)$.   ⊣

Case 2: $B \notin M$

We want to determine if every GC-model of (K, M) entails $B(a)$. To achieve that, we navigate bottom up in the configuration space which is essentially the space of possible individual allocations and ground extensions to the predicates in M. Let $\mathcal{G}_{(K,M)} = \{G_1, ..., G_\lambda\}$, where $G_1 < G_2 < ... < G_\lambda$.

IC Algorithm:

1] Initiate Stack $:= \mathcal{G}_{(K,M)}$.
2] for $i = 1$ to λ
3]   If $G_i \in$ Stack:
4]   Check K_{G_i} for satisfiability.
5]   If YES:
6]   Check $K_{G_i}^+$ for satisfiability.
7]   If YES return FALSE.
8]   Else remove all $G_j \succ G_i$ from Stack.
9] return TRUE.

That the above algorithm terminates is obvious, because there is only one loop. Moreover the command in line 9, outside of the loop, guarantees that it will return either TRUE or FALSE.

Proposition 3. The IC algorithm returns TRUE if and only if (K, M) entails $B(a)$. ⊣

To demonstrate how this whole procedure works, we give a simple example.

Example. Let K be a knowledge base consisting of the following axioms:

$$B(a), \neg B(b), r(b, c), \rho(a, b), \rho(a, c), \exists r.\neg A \sqsubseteq A$$

Let $M = \{A\}$. Then $Part(Ind(K)) = \{I_1, I_2, I_3, I_4, I_5\}$ where

$$I_1 = \{\{a\}, \{b\}, \{c\}\}$$
$$I_2 = \{\{a, b\}, \{c\}\}$$
$$I_3 = \{\{a, c\}, \{b\}\}$$
$$I_4 = \{\{a\}, \{b, c\}\}$$
$$I_5 = \{\{a, b, c\}\}.$$

Figure 1 is a visualization of our configuration space. Each possible individual allocation corresponds to a lattice of possible ground extensions for the closed predicates, which in our case consists of just A. The restriction of the search space to I-complete sets of possible extensions, with respect to an individual allocation I, is portrayed by the apparent reduction of points in I_2-I_5.

Suppose that we want to check the assertion $\neg(A \sqcap \forall \rho.A)(a)$ for entailment. This basically means that not all individuals can be interpreted as members of the extension of A. Then the IC algorithm will look bottom-up for grounded models of K wrt M. If a model is found, then an augmented knowledge base will be built, consisting of the current pointwise restriction and the negation of the given assertion, which in our case is just $(A \sqcap \forall.\rho A)(a)$. In case this augmented KB is found to be satisfiable, the algorithm will halt, giving FALSE as an answer. Otherwise it will remove from the Stack all points which are above, hence reducing further the remaining exploration. For this particular instance,

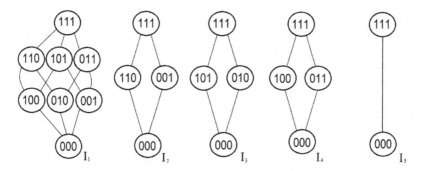

Fig. 1. The configuration space of (K, M).

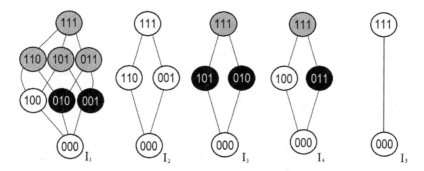

Fig. 2. The distribution of GC-models of (K, M) in the configuration space.

(K, M) entails $\neg(A \sqcap \forall.\rho A)(a)$, so no minimal model which satisfies $(A \sqcap \forall.\rho A)(a)$ can be found, and so the algorithm will return TRUE.

Note that the entailment holds exactly because of the minimality, i.e. there are grounded models where all individuals belong to A. Figure 2 gives an account of the distribution of grounded models and GC-models of (K, M) over the configuration space. Points in white are those that do not correspond to any model of K_M, points in black correspond to GC-models and points in grey to the rest of the grounded models. All the points in grey are exactly those that will be never "visited", i.e. at some step they will be removed from the Stack.

Complexity and Optimization Considerations. Considering that the presented algorithm requires at most exponentially many calls of the $\mathcal{ALCO}$ reasoner, each of which requires exponential time, we get that the overall complexity is still in ExpTime. The lower bound is ExpTime as well as in the case $M = \emptyset$ the inference problem turns into standard reasoning in $\mathcal{ALCO}$ which is known to be ExpTime-complete. For more expressive description logics, the complexity of the black-box reasoning part will dominate and hence determine the overall complexity.

Regarding the practical runtime behavior, we expect a significant improvement through the removal of points that results from the command in line 8.

That is because the algorithm, in accordance with the defined linear order, will try smaller points of the configuration space first and once a model is found, the algorithm will stop looking at the rest of the branch.

Of course there is room for optimization of this algorithm. Notably from the example we can see how two out of the five lattices should have been rejected from the start, since they represent individual allocations which are incompatible with the given knowledge base. More thoroughly, one could remove points which correspond to assignments which are not consistent with the axioms in the knowledge base.

On the other hand, the results we have acquired so far can be directly extended to more complex languages. That follows from the fact that in none of the proofs supporting this study did we rely on the limitations of $\mathcal{ALCO}$. In effect, we have used the constructive capabilities of our language, in creating new knowledge bases that represent the notion of grounding and different points in the configuration space. But we have not appealed to any restrictions imposed by the specific syntax of $\mathcal{ALCO}$, with the exception of course of the property of decidabilty, which is implicit wherever a decision procedure is regarded.

5 Minimality Checking: A Non-standard Reasoning Task

In this section we present a solution to the task of determining whether a specific grounded model is minimal by calling the standard DL reasoner just once. It can be of use in devising algorithms for other reasoning problems in grounded circumscription, but also maybe in some optimized variant of the IC algorithm presented previously.

Definition 10. Let K_M be a grounded KB where $M \cap N_C = \{A_1, ..., A_n\}$ and $M \cap N_r = \{r_1, ..., r_m\}$ and let $\mathcal{I}$ be a model of K_M. We call *down-the-chain axioms* with respect to $\mathcal{I}$, the following set of GCIs:

   I. $\{a\} \equiv Ext^{\mathcal{I}}(\{a\})$ for every $a \in Ind(K)$,

   II. $Ext^{\mathcal{I}}(\neg\{a\}) \sqsubseteq \neg\{a\}$ for every $a \in Ind(K)$,

   III. $A_i \sqsubseteq Ext^{\mathcal{I}}(A_i)$ for every $i \in \{1, ..., n\}$,

   IV. $B_{(a,j)} \equiv \{c \in Ind(K) | (a^{\mathcal{I}}, c^{\mathcal{I}}) \in r_j^{\mathcal{I}}\}$ for every $a \in Ind(K)$ and $j \in \{1, ..., m\}$,

   V. $\{a\} \sqsubseteq \forall r_j.B_{(a,j)}$ for every $a \in Ind(K)$ and $j \in \{1, ..., m\}$,

   VI. $\top \sqsubseteq \exists r.\left(\left(\bigsqcup_{i \in \{1,...,n\}} \left(Ext^{\mathcal{I}}(A_i) \sqcap \neg A_i\right)\right) \sqcup \left(\bigsqcup_{\substack{j \in \{1,...,m\} \\ a \in Ind(K) \\ c \in B_{(a,j)}}} \left(\{a\} \sqcap \forall r_j.\neg\{c\}\right)\right)\right)$,

where r is a fresh role, i.e. it does not appear in K. K_M augmented with the down-the-chain axioms with respect to a model is called a *confining* of K_M and symbolized $K_M^{\mathcal{I}-}$, where $\mathcal{I}$ is the respective model.   ⊣

Notice that the number of axioms in each of the categories I-V depends on M whereas VI is one single axiom. The next lemma shows how we can find a smaller grounded model than a given one, if there exists one. Intuitively this is like going down in the lattice of possible grounded models, hence the terminology.

Lemma 5. Let (K, M) be a *GC-$\mathcal{ALCO}$-KB* and let $\mathcal{I}$ be a model of K_M. There exists a model $\mathcal{J}$ of K_M such that $\mathcal{J} \prec_M \mathcal{I}$ if and only if $K_M^{\mathcal{I}^-}$ is satisfiable. ⊣

For direct practical use, the above lemma is more conveniently expressed in the following form:

Corollary 1. *(Minimality Check)* Let (K, M) be a *GC-$\mathcal{ALCO}$-KB* and let $\mathcal{I}$ be a model of K_M. If $K_M^{\mathcal{I}^-}$ is unsatisfiable, then $\mathcal{I}$ is a GC-model of (K, M). ⊣

6 Conclusions

In our paper we have refined and rectified the foundational definition of grounded circumscription and have produced some first results as a basis for further research. Starting from a definition that is more accurate in incorporating the intuition behind grounded circuscription, we have an improved solution to the satisfiability task which now does not require any non-standard description logic and can be solved by a single call to an off-the-shelf DL reasoner. Moreover, we have provided an algorithm for instance checking, which was only insufficiently covered in the original paper on grounded circumscription.

Apart from the algorithm itself, the theory provided gives a well-founded understanding of the general potential of grounded circumscription, as redefined here. The configuration space can prove to be a useful notion for devising other non-standard reasoning algorithms. The down-the-chain axioms and minimality check as a sub-task could contribute to solving other reasoning tasks within grounded circumscription as well.

As mentioned earlier, an advantage of our approach is that all our results hold if $\mathcal{ALCO}$ is replaced by a more complex language, as long as it is decidable. Certainly there is a lot of space for further development of grounded circumscription. It remains to be seen whether the IC algorithm performs well in practice and/or can be sufficiently optimized further.

One of our main aims was to reduce as much of the reasoning as possible to standard DL reasoning. This is achieved, in our opinion to the largest extent possible. With this feature, our theory is implementation-friendly, and one main future objective is to create a reasoner for grounded circumscription, which will of course be working on top of an efficient standard DL reasoner.

References

1. Baader, F., Calvanese, D., McGuinness, D.L., Nardi, D., Patel-Schneider, P.F. (eds.): The Description Logic Handbook: Theory Implementation and Applications. Cambridge University Press, New York (2003)
2. Bonatti, P.A., Lutz, C., Wolter, F.: The Complexity of Circumscription in DLs (2014). CoRR abs/1401.3476
3. Delivorias, S., Rudolph., S.: Grounded Circumscription in Description Logics (2015)
4. Hitzler, P., Krötzsch, M., Rudolph, S.: Foundations of Semantic Web Technologies. Chapman & Hall/CRC, Boca Raton (2009)

5. McCarthy, J.: Circumscription – a form of non-monotonic reasoning. Artif. Intell. **13**, 27–39 (1980)
6. Rudolph, S.: Foundations of description logics. In: Polleres, A., d'Amato, C., Arenas, M., Handschuh, S., Kroner, P., Ossowski, S., Patel-Schneider, P. (eds.) Reasoning Web 2011. LNCS, vol. 6848, pp. 76–136. Springer, Heidelberg (2011)
7. Sengupta, K., Krisnadhi, A.A., Hitzler, P.: Local closed world semantics: grounded circumscription for OWL. In: Aroyo, L., Welty, C., Alani, H., Taylor, J., Bernstein, A., Kagal, L., Noy, N., Blomqvist, E. (eds.) ISWC 2011, Part I. LNCS, vol. 7031, pp. 617–632. Springer, Heidelberg (2011)
8. W3C OWL Working Group. OWL 2 Web Ontology Language: Document Overview (2009). http://www.w3.org/TR/owl2-overview/

Query Rewriting under Linear
$\mathcal{EL}$ Knowledge Bases

Mirko M. Dimartino[1]([✉]), Andrea Calì[1,2], Alexandra Poulovassilis[1],
and Peter T. Wood[1]

[1] Knowledge Lab, Birkbeck, University of London, London, UK
{mirko,andrea,ap,ptw}@dcs.bbk.ac.uk
[2] Oxford-Man Institute of Quantitative Finance,
University of Oxford, Oxford, UK

Abstract. With the adoption of the recent SPARQL 1.1 standard, RDF databases are capable of directly answering more expressive queries than simple conjunctive queries. In this paper we exploit such capabilities to answer conjunctive queries (CQs) under ontologies expressed in the description logic called *linear $\mathcal{EL}^{\ell in}$*, a restricted form of $\mathcal{EL}$. In particular, we show a query answering algorithm that rewrites a given CQ into a conjunctive regular path query (CRPQ) which, evaluated on the given instance, returns the correct answer. Our technique is based on the representation of infinite unions of CQs by non-deterministic finite-state automata. Our results achieve optimal data complexity, as well as producing rewritings straightforwardly implementable in SPARQL 1.1.

1   Introduction

Ontologies have been successfully employed in conceptual modelling of data in several areas, especially Information Integration and the Semantic Web. An ontology is a specification of the domain of interest of an application, and it is usually specified in terms of logical rules which on the one hand restrict the form of the underlying data, and on the other hand allow for *inference* of information that is not explicitly contained in the data. Description Logic (DL) is a common family of knowledge representation formalisms that are able to capture a wide range of ontological constructs [2]; they are based on *concepts* (unary predicates representing classes of individuals) and *roles* (binary predicates representing relations between classes). A DL knowledge base consists of a TBox (*terminological* component) and an ABox (*assertional* component); the former is a conceptual representation of the schema, while the latter is an instance of the schema. It is important to note that a usual assumption in this context is the so-called *open-world* assumption, that is, the information in the ABox is sound but not complete; the TBox, in particular, determines how the ABox is to be completed with additional information so as to answer queries. Answers to a query in this context are called, following database parlance, *certain answers*, as they correspond to the answers that are true in all models of the theory constituted by the knowledge base. This corresponds to *cautious* reasoning as opposed to *bold* reasoning, where in the latter an answer is returned if it is entailed by

© Springer International Publishing Switzerland 2016
M. Ortiz and S. Schlobach (Eds.): RR 2016, LNCS 9898, pp. 61–76, 2016.
DOI: 10.1007/978-3-319-45276-0_6

at least one model. The set of all models (which is not necessarily finite) is represented by the so-called *expansion* (called *chase* in database parlance) of an ABox $\mathcal{A}$ according to a TBox $\mathcal{T}$; this is illustrated in the following example.

Example 1. Consider the TBox $\mathcal{T}$ constituted by the assertions $C \sqsubseteq A$ and $A \sqsubseteq \exists S.C$. The concept $\exists S.C$ denotes the objects connected via the role S to some object belonging to the concept C; in other words, it contains all x such that $S(x, y)$ and $C(y)$ for some y. The first assertion means that every object in the class C is also in A; the second means that every object in the class A is also in the class represented by $\exists S.C$. Now suppose we have the ABox $\mathcal{A} = \{A(a)\}$; we can *expand* $\mathcal{A}$ according to the TBox $\mathcal{T}$ so as to add to it all atoms entailed by $(\mathcal{T}, \mathcal{A})$; we therefore add $S(a, z_0)$ and $C(z_0)$, where z_0 is a so-called *labelled null*, that is, a placeholder for an unknown value of which we know the existence. Given the query q defined as $q(x) \leftarrow S(x, y)$, the answer to it under $(\mathcal{T}, \mathcal{A})$ is $\{a\}$ because $S(a, z_0)$ is entailed by $(\mathcal{T}, \mathcal{A})$; in fact, the certain answers to q are obtained by evaluating q on the expansion and by considering answers that do not contain nulls. If we consider the query q_1 defined as $q_1(x) \leftarrow C(x)$, the answer is empty because z_0, though known to exist, is not known.

Answers to queries over DL knowledge bases can be computed, in certain cases, by a technique called *query rewriting*. In query rewriting, starting from a given query q, a new query q' is computed according to a knowledge base $\mathcal{K} = (\mathcal{T}, \mathcal{A})$, such that the answers to q on $\mathcal{K}$ are obtained by evaluating q' on $\mathcal{A}$ only; it is said that q is *rewritten* into q' and that q' is the *perfect rewriting of q with respect to $\mathcal{T}$*. The language of q', called the *target language*, can be more expressive than that of q. A common rewriting technique for DLs and other knowledge representation formalisms, inspired by resolution in Logic Programming, has *union of conjunctive queries* as target language.

Example 2. Let us consider again the knowledge base of Example 1. The query q is rewritten into the query q' defined as $q(x) \leftarrow A(x) \cup S(x, y)$; intuitively, q' captures the fact that, to search for objects from which some other object is connected via the role S, we need also to consider objects in A, because the TBox might infer the former from the latter objects. The evaluation of q' on $\mathcal{A}$ returns the correct answer.

In this paper we consider a DL which we call $\mathcal{EL}^{\ell in}$ [17]. When executed on $\mathcal{EL}^{\ell in}$ TBoxes, the above rewriting technique does not guarantee termination. We therefore resort to a more expressive target language for the rewriting, namely *conjunctive regular path queries* (CRPQs).

Example 3. Consider the TBox $\mathcal{T} = \{\exists R.A \sqsubseteq A\}$ and the query q defined as $q(x) \leftarrow A(x)$. It is easy to see that the above rewriting technique produces an infinite union of conjunctive queries: $q(x) \leftarrow A(x)$, $q(x) \leftarrow R(x, y), A(y)$ and all conjunctive queries of the form $q(x) \leftarrow R(x, y_1), \ldots, R(y_k, y_{k+1}), A(y_{k+1})$, with $k \geqslant 1$. Now, in order to capture this infinite rewriting, we can resort to the CRPQ rewriting q' defined as $q(x) \leftarrow R^*(x, y), A(y)$.

In this paper we propose a novel rewriting technique for the DL $\mathcal{EL}^{\ell in}$, where the query language is that of *conjunctive queries* (CQs) and the target language is that of CRPQs. This allows us to devise a query answering algorithm that has optimum asymptotic complexity and relies on pure rewriting, without ABox expansion. Notice that rewriting is generally considered efficient as the processing operates solely on the query, while the ABox, which is normally considered to be much larger than the query and the TBox, comes into play only at the last step, when the rewriting is evaluated on it.

Our contributions are as follows.

- For illustrative purposes and technical reasons, we show a rewriting algorithm for conjunctive queries on $\mathcal{EL}^{\ell in}$ knowledge bases, which relies on a resolution-like procedure widely adopted in the literature (see e.g. [10]).
- We present a novel rewriting technique, based on non-deterministic finite-state automata, for *atomic* queries on $\mathcal{EL}^{\ell in}$ knowledge bases, with CRPQs as target language. Intuitively, the expressive power of CRPQs is able to finitely capture the infinite rewriting branches of the above algorithm.
- Finally, based on the rewriting technique for atomic queries, we present a technique for rewriting CQs into CRPQs. This is achieved by splitting the problem in two: first we deal with assertions that do not introduce labelled nulls; then, we show that the rest of the assertions are guaranteed to have only tree-like (or, more precisely, forest-like) models; this allows us to capture all paths (including the infinite ones) from roots in the forest by means of finite-state automata. The final rewriting is a CRPQ whose evaluation on the given ABox returns the correct answers to the initial CQ. Since CRPQs can be straightforwardly expressed in SPARQL 1.1, by the means of property paths, our approach is suitable for real-world settings.
- Our technique achieves optimal computational cost in *data complexity*, that is where the TBox and the query are fixed and the ABox alone is a variable input; in fact, our algorithm runs in NLOGSPACE in data complexity, which is the known (tight) bound for CQ answering under $\mathcal{EL}^{\ell in}$ knowledge bases. Notice also that, regarding *combined complexity* (where query, TBox and ABox all constitute the variable input), our rewriting is expressed in the language of CRPQs, which can be evaluated in NP. Moreover, in the case of "simple" queries such as atomic queries, our rewriting is expressed as a regular path query, which can be evaluated in NLOGSPACE.

2 Preliminaries

In this section we present the formal notions which will be used in the rest of the paper.

2.1 Description Logics

We briefly introduce the syntax of the $\mathcal{EL}^{\ell in}$ description logic (DL) [16,18]. The alphabet contains three pairwise disjoint and countably infinite sets of *concept names* A, *role names* R, and *individual names* I. A *complex concept* C is constructed from a special primitive concept $\top$ ('top'), concept names and role names using the following grammar: $C ::= A \mid \exists R.C \mid \exists R.\top$, where $A \in$ A and $R \in$ R. The set of complex concepts is denoted by C. A *terminological box* (or TBox) $\mathcal{T}$ is a finite set of *concept* and *role inclusion axioms* of the form $C_1 \sqsubseteq C_2$ and $R_1 \sqsubseteq R_2$, where $C_1, C_2 \in$ C and $R_1, R_2 \in$ R. An *assertion box* (or ABox) $\mathcal{A}$ is a finite set of *concept* and *role assertions* of the form $C(a)$ and $R(a,b)$, where $C \in$ C, $R \in$ R and $a, b \in$ I. Given an ABox $\mathcal{A}$, we denote by $\text{ind}(\mathcal{A})$ the set of individual names that occur in $\mathcal{A}$. Taken together, $\mathcal{T}$ and $\mathcal{A}$ comprise a *knowledge base* (or KB) $\mathcal{K} = (\mathcal{T}, \mathcal{A})$.

We adopt the semantics of DL defined in terms of interpretations. An *interpretation* $\mathcal{I}$ is a pair $(\Delta^{\mathcal{I}}, \cdot^{\mathcal{I}})$ that consists of a non-empty *domain of interpretation* $\Delta^{\mathcal{I}}$ and an *interpretation function* $\cdot^{\mathcal{I}}$ which assigns (i) an element $a^{\mathcal{I}} \in \Delta^{\mathcal{I}}$ to each individual name a, (ii) a subset $A^{\mathcal{I}} \subseteq \Delta^{\mathcal{I}}$ to each concept name A and (iii) a binary relation $P^{\mathcal{I}} \subseteq \Delta^{\mathcal{I}} \times \Delta^{\mathcal{I}}$ to each role name P. We adopt the *unique name assumption* (UNA); therefore distinct individuals are assumed to be interpreted by distinct domain elements. The interpretation function $\cdot^{\mathcal{I}}$ is extended inductively for complex concepts by taking:

$$(\exists R.\top)^{\mathcal{I}} = \{u \mid \text{there is a } v \text{ such that } (u,v) \in R^{\mathcal{I}}\},$$
$$(\exists R.C)^{\mathcal{I}} = \{u \mid \text{there is a } v \in C^{\mathcal{I}} \text{ such that } (u,v) \in R^{\mathcal{I}}\}.$$

We now define the satisfaction relation $\models$ for inclusions and assertions:

$$\mathcal{I} \models C_1 \sqsubseteq C_2 \text{ if and only if } C_1^{\mathcal{I}} \subseteq C_2^{\mathcal{I}},$$
$$\mathcal{I} \models R_1 \sqsubseteq R_2 \text{ if and only if } R_1^{\mathcal{I}} \subseteq R_2^{\mathcal{I}},$$
$$\mathcal{I} \models C(a) \text{ if and only if } a^{\mathcal{I}} \in C^{\mathcal{I}},$$
$$\mathcal{I} \models R(a,b) \text{ if and only if } (a^{\mathcal{I}}, b^{\mathcal{I}}) \in R^{\mathcal{I}}.$$

We say that an interpretation $\mathcal{I}$ is a *model* of a knowledge base $\mathcal{K} = (\mathcal{T}, \mathcal{A})$, written $\mathcal{I} \models \mathcal{K}$, if it satisfies all concept and role inclusions of $\mathcal{T}$ and all concept and role assertions of $\mathcal{A}$. A TBox is said to be in *normal form* if each of its concept inclusion axioms is of the forms $A \sqsubseteq B$, $\exists P.C \sqsubseteq A$, or $A \sqsubseteq \exists P.C$, where $A, B \in$ A, $C \in$ A $\cup \{\top\}$ and $P \in$ R. We recall that every $\mathcal{EL}$ TBox can be transformed into an equivalent TBox in normal form of size that is linear in the size of the original TBox [14].

2.2 Regular Languages and Conjunctive Regular Path Queries

We assume the reader is familiar with regular languages, represented either by regular expressions or nondeterministic finite state automata. A *nondeterministic finite state automaton* (NFA) over a set of symbols Σ is a tuple $\alpha = (Q, \Sigma, \delta, q_0, F)$, where Q is a finite set of *states*, $\delta \subseteq Q \times \Sigma \times Q$ the *transition relation*, $q_0 \in Q$ the *initial state*, and $F \subseteq S$ the set of *final states*. We use $L(\alpha)$ to denote the regular language defined by an NFA α, and $(\Sigma)^*$ to denote the set of all strings over symbols in Σ, including the *empty string* ϵ.

In order to define queries, we also need to assume the existence of a countably infinite set of *variables* V. A *term* t is an individual name in I or a variable in V. An *atom* is of the form $\alpha(t, t')$, where t, t' are terms, and α is an NFA or regular expression defining a regular language over $R \cup A$. We say that a string $s \in (R \cup A)^*$ is a *path*.

A *conjunctive regular path query* (CRPQ) $\boldsymbol{q}$ of arity n has the form $q(\boldsymbol{x}) \leftarrow \gamma(\boldsymbol{x}, \boldsymbol{y})$, where $\boldsymbol{x} = x_1, \ldots, x_n$ and $\boldsymbol{y} = y_1, \ldots, y_m$ are variables, and $\gamma(\boldsymbol{x}, \boldsymbol{y})$ is a set of atoms with variables from $\boldsymbol{x}$ and $\boldsymbol{y}$. $q(\boldsymbol{x})$ is called the *head* of $\boldsymbol{q}$ and is denoted by $head(\boldsymbol{q})$, and $\gamma(\boldsymbol{x}, \boldsymbol{y})$ is the *body* of $\boldsymbol{q}$ and denoted by $body(\boldsymbol{q})$. The variables in $\boldsymbol{x} = x_1, \ldots, x_n$ are the *answer variables* of $\boldsymbol{q}$, while those in $\boldsymbol{y} = y_1, \ldots, y_m$ are the *existentially quantified variables* of $\boldsymbol{q}$. A *Boolean CRPQ* is a CRPQ with no answer variables. A *regular path query* (RPQ) is a CRPQ with a single atom in its body. A *path query* (PQ) is an RPQ $q = head(q) \leftarrow \alpha(x, y)$ such that $\alpha \in (R \cup A)^*$, where α is the *path* of q denoted by $path(q)$.

A *conjunctive query* (CQ) $\boldsymbol{q}$ is a CRPQ such that, for each atom $\alpha(t, t') \in body(\boldsymbol{q})$, $\alpha \in (R \cup A)$. Informally, CQs disallow regular expressions in their bodies. Given a CRPQ $\boldsymbol{q}$ with answer variables $\boldsymbol{x} = x_1, \ldots, x_n$ and an n-tuple of individuals $\boldsymbol{a} = (a_1, \ldots, a_n)$, we use $\boldsymbol{q}(\boldsymbol{a})$ to refer to the Boolean query obtained from $\boldsymbol{q}$ by replacing x_i with a_i in $body(\boldsymbol{q})$, for every $1 \leqslant i \leqslant n$.

We now define the semantics of CRPQs [7]. Given the individual names a, b, an interpretation $\mathcal{I}$, and a regular language α over the alphabet $R \cup A$, we have that $\mathcal{I} \models a \xrightarrow{\alpha} b$ if and only if there is some $w = u_1 \ldots u_n \in L(\alpha)$ and some sequence $e_0, \ldots, e_n$ with $e_i \in \Delta^{\mathcal{I}}$, $0 \leqslant i \leqslant n$, such that $e_0 = a^{\mathcal{I}}$ and $e_n = b^{\mathcal{I}}$, and for all $1 \leqslant i \leqslant n$: *(i)* if $u_i = A \in A$, then $e_{i-1} = e_i \in A^{\mathcal{I}}$; *(ii)* if $u_i = R \in R$, then $(e_{i-1}, e_i) \in R^{\mathcal{I}}$. A *match* for a Boolean CRPQ $\boldsymbol{q}$ in an interpretation $\mathcal{I}$ is a mapping π from the terms in $body(\boldsymbol{q})$ to the elements in I such that: *(1)* $\pi(c) = c$ if $c \in$ I; *(2)* $\mathcal{I} \models \pi(t) \xrightarrow{\alpha} \pi(t')$ for each atom $\alpha(t, t')$ in $\boldsymbol{q}$.

Note that, to avoid notational clutter, we do not allow unary atoms in the body of the query. In fact, each atom of the form $A(t)$, where $A \in A$ and $t \in V \cup I$, can be always replaced by a binary atom $A(t, z)$, where z is a variable. However, for better legibility, we use unary atoms in some examples throughout the paper. We write $\mathcal{I} \models \boldsymbol{q}$ if there is a match for $\boldsymbol{q}$ in $\mathcal{I}$, and $\mathcal{K} \models \boldsymbol{q}$ if $\mathcal{I} \models \boldsymbol{q}$ for every model $\mathcal{I}$ of the KB $\mathcal{K}$. For brevity, given an ABox $\mathcal{A}$ we use $\mathcal{A} \models \boldsymbol{q}$ to refer to $(\varnothing, \mathcal{A}) \models \boldsymbol{q}$, where $(\varnothing, \mathcal{A})$ is a knowledge base with empty TBox. Given a CRPQ $\boldsymbol{q}$ of arity n we say that a tuple of individual names $\boldsymbol{a} = (a_1, \ldots, a_n)$ is a *certain answer* for $\boldsymbol{q}$ with respect to a KB $\mathcal{K}$ if and only if $\mathcal{K} \models \boldsymbol{q}(\boldsymbol{a})$.

3   Rewriting of Conjunctive Queries into First-Order Queries

In this section we show a technique for rewriting CQs into a *union of conjunctive queries* under an $\mathcal{EL}^{\ell in}$ TBox. We base our approach on the rewriting algorithm proposed in [9], which deals with *DL-Lite$_{\mathcal{R}}$* KBs. We do this to establish correctness of the rewriting approach with the NFAs illustrated in Sect. 4. The technique is based on two steps: a *reduction* step, which eliminates atoms that are more specific than some other atom, and the actual *rewriting* step, which is similar to the resolution step in logic programming. Notice that the algorithm might not terminate; we present it for technical reasons, as in Sects. 4 and 5 we will show that our CRPQ rewriting captures all the rewriting branches produced by the algorithm, including infinite ones.

Following the approach of [9], we say that a term of an atom in a query is *bound* if it corresponds to (i) an answer variable, (ii) a shared variable, that is, a variable occurring at least twice in the query body, or (iii) a constant, that is an element in I. Conversely, a term of an atom in a query is *unbound* if it corresponds to a non-shared existentially quantified variable. As usual, we use the symbol '_' to represent an unbound term.

A set of atoms $A = \{a_1, \ldots, a_n\}$, where $n > 2$, *unifies* if there exists a substitution ϕ, called *unifier* for A, such that (i) if $t \in$ I, then $\phi(t) = t$, and (ii) $\phi(a_1) = \cdots = \phi(a_n)$. *Reduce* is a function that takes as input a conjunctive query q and a set of atoms S occurring in the body of q and returns a conjunctive query q obtained by applying to q the *most general unifier* between the atoms of S. We point out that, in unifying a set of atoms, each occurrence of the _ symbol is considered to be a different unbound variable.

We now define when concept and role inclusion axioms are *applicable* to atoms in a query. An axiom I is *applicable to an atom* $A(x_1, x_2)$ for $A \in$ A if I is of the form $B \sqsubseteq A$, $\exists R.\top \sqsubseteq A$ or $\exists R.B \sqsubseteq A$. An axiom I is *applicable to an atom* $P(x_1, x_2)$ for $P \in$ R if (1) $x_2 = \_$ and the right-hand side of I is $\exists P.\top$ or $\exists P.A$; or (2) the right-hand side of I is P. An axiom I is *applicable to a pair of atoms* $P(x_1, x_2), A(x_2, x_3)$ if x_2 does not appear in other atoms of the query body, $x_3 = \_$ and I is of the form $C \sqsubseteq \exists P.A$. Below we define the set of rewriting rules for atoms in the query body. Let I be an inclusion assertion that is applicable to a sequence of query atoms g. The sequence of atoms obtained from g by applying I, denoted by $gr(g, I)$, is defined as follows:

(a) If $g = A(x_1, x_2)$ and $I = B \sqsubseteq A$, then $gr(g, I) = B(x_1, x_2)$;
(b) If $g = A(x_1, x_2)$ and $I = \exists P.\top \sqsubseteq A$, then $gr(g, I) = P(x_1, \_)$;
(c) If $g = A(x_1, x_2)$ and $I = \exists P.B \sqsubseteq A$, then $gr(g, I) = P(x_1, z_1), B(z_1, \_)$, where z_1 is a fresh variable;
(d) If $g = P(x_1, \_)$ and $I = A \sqsubseteq \exists P.\top$ or $I = A \sqsubseteq \exists P.B$, then $gr(g, I) = A(x_1, \_)$;
(e) If $g = P(x_1, x_2)$ and $I = R \sqsubseteq P$, then $gr(g, I) = R(x_1, x_2)$;
(f) If $g = P(x_1, x_2), A(x_2, \_)$ and $I = C \sqsubseteq \exists P.A$, then $gr(g, I) = C(x_1, \_)$;

We denote by $\mathsf{Rewrite}(q, \mathcal{T})$ the rewriting procedure that generates the perfect rewriting of q with respect to $\mathcal{T}$ (see Fig. 1).

Example 4. Consider applying the Rewrite procedure to a query q of the form $q(x) \leftarrow R(x, y), R(\_, y)$ over the TBox $\{A \sqsubseteq \exists R.\top\}$, where $A \in$ A and $R \in$ R. In this query, the atoms $R(x, y)$ and $R(\_, y)$ unify, and executing $\mathsf{Reduce}(q, \{R(x, y), R(\_, y)\})$ yields the atom $R(x, y)$. The variable y is now unbound, so can be replaced by "_" (a *don't care*). Note that the reduction step produces a query marked with '0' whilst the rewriting step marks queries with '1', and only queries marked with '1' are added to the output set. We adopt

---

Algorithm 1. Algorithm Rewrite$(q, \mathcal{T})$

Data: Conjunctive query q, TBox $\mathcal{T}$.
Result: Union of conjunctive queries Q.
$Q := \{\langle q, 1 \rangle\}$;
repeat
$\quad Q' := Q$;
$\quad$ **foreach** $\langle qr, x \rangle \in Q'$ **do**
$\qquad$ /* Reduction step   */
$\qquad$ **if** *there exists* $I \in \mathcal{T}$ *such that* I *is not applicable to* qr **then**
$\qquad\quad$ **foreach** *set of atoms* $S \subseteq body(qr)$ **do**
$\qquad\qquad$ **if** S *unify* **then**
$\qquad\qquad\quad$ $Q := Q \cup \langle \mathsf{Reduce}(qr, S), 0 \rangle$

$\qquad$ /* Rewriting step   */
$\qquad$ **foreach** *axiom* $I \in \mathcal{T}$ **do**
$\qquad\quad$ **if** I *is applicable to* qr **then**
$\qquad\qquad$ $qr' :=$ rewrite qr according to I ;
$\qquad\qquad$ $Q := Q \cup \langle qr', 1 \rangle$

until $Q' = Q$;
$Q_{fin} := \{q \mid \langle q, 1 \rangle \in Q\}$;
return Q_{fin}

---

this approach to avoid redundancy in the output set, as a query marked with '0' is always contained in a query marked with '1'. Now, the axiom $\{A \sqsubseteq \exists R.\top\}$ can be applied to $R(x, \_)$, whereas, before the reduction process, it could not be applied to any atom of the query. Following this, the rewriting step reformulates the query to $q(x) \leftarrow A(x, \_)$ which is added to the output set. For more details on the rewriting procedure refer to [9,10].

Now we show that each disjunct of the perfect rewriting of an atomic concept query with respect to an $\mathcal{EL}^{\ell in}$ TBox is of a special form called a *simple path conjunctive query*, defined below. We then define some technical lemmas which will be used in Sect. 4 for the rewriting of atomic concepts by means of a finite-state automaton.

Definition 1. *A conjunctive query q is a simple path conjunctive query (SPCQ) if $body(q)$ is of one of the following forms: (i) $A(x_1, x_2)$;*
(ii) $P_1(x_1, y_1), P_2(y_1, y_2), \ldots, P_{n-1}(y_{n-2}, y_{n-1}), P_n(y_{n-1}, x_2)$; or
(iii) $P_1(x_1, y_1), P_2(y_1, y_2), \ldots, P_{n-1}(y_{n-2}, y_{n-1}), P_n(y_{n-1}, y_n), A(y_n, x_2)$, where:
x_1, x_2 are terms; for each i, y_i is an existentially quantified variable and $y_i \neq y_{i+1}$; $n \geqslant 1$; $A \in \mathsf{A}$ and $P_1, \ldots, P_n \in \mathsf{R}$.

Note that query q in Example 4 is not an SPCQ. An SPCQ $head(q) \leftarrow Z_1(x_0, x_1), \ldots, Z_n(x_{n-1}, x_n)$ is equivalent to an RPQ of the form $head(q) \leftarrow Z_1 \ldots Z_n(x_0, x_n)$; thus, throughout the paper we will use either the RPQ form or the CQ form of a SPCQ, whichever is more natural in the given context.

For instance, given a SPCQ q, with a little abuse of notation we have that $path(q)$ is $Z_1 \ldots Z_n$.

Given two paths p, p', we say that p' *contains* p, written $p \sqsubseteq p'$, if for each ABox $\mathcal{A}$ and for each tuple $\boldsymbol{a} = (a, a')$ it holds that, if $\mathcal{A} \models q() \leftarrow p(\boldsymbol{a})$ then $\mathcal{A} \models q() \leftarrow p'(\boldsymbol{a})$. Given a path p and an NFA $\mathcal{N}$ over $(R \cup A)^*$ we say that $\mathcal{N}$ *contains* p, written $p \sqsubseteq \mathcal{N}$ if there exists some $\alpha \in L(\mathcal{N})$ such that $p \sqsubseteq \alpha$.

Lemma 1. *Given an SPCQ q, an $\mathcal{EL}^{lin}$ TBox $\mathcal{T}$ and an axiom $\rho \in \mathcal{T}$ that is not applicable to $body(q)$, for each set of atoms $S \subseteq body(q)$ that unify, ρ is also not applicable to $body(Reduce(q, S))$.*

Proof (Sketch). We consider all the possible cases of ρ.

Case 1: ρ is of the type $B \sqsubseteq A$, $\exists P.\top$, $\exists R.B \sqsubseteq A$, $R \sqsubseteq P$. ρ is not applicable to $body(q)$ if an atom $A(x_1, x_2)$ or $P(x_1, x_2)$ are not in $body(q)$. If $A(x_1, x_2)$ or $P(x_1, x_2)$ is not in $body(q)$, then $A(x_1, x_2)$ or $P(x_1, x_2)$ is clearly not in $body(Reduce(q, S))$ and the claim follows.

Case 2: ρ is of the type $A \sqsubseteq \exists R.\top$. ρ is not applicable to $body(q)$ if an atom $R(x_1, x_2)$ is not in $body(q)$, or if $R(x_1, x_2)$ is in $body(q)$ and $x_2 \neq \_$. If $R(x_1, x_2)$ is not in $body(q)$, then $R(x_1, x_2)$ is clearly not in $body(Reduce(q, S))$. If $R(x_1, x_2)$ is in $body(q)$ and $x_2 \neq \_$, since q is an SPCQ, this happens only if $R(x_1, x_2)$ is not the last atom of the path. The only way to have $x_2 = \_$ is to unify $R(x_1, x_2)$ with the atom at its right as it is the only atom that can have x_2. If there exists a unification, the reduction produces an atom $R(x_i, x_i)$, with $x_i \neq \_$ and the claim follows.

Case 3: ρ is of the type $A \sqsubseteq \exists R.B$. Since q is an SPCQ, ρ is not applicable to $body(q)$ if the atoms $R(x_1, x_2), B(x_2, x_3)$ are not in $body(q)$ or if $R(x_1, x_2)$ is in $body(q)$ and $x_2 \neq \_$. It is easy to see that this case is similar to the case where ρ is of the type $A \sqsubseteq \exists R.\top$.

Following from Lemma 1 we have that, to rewrite CQs with respect to $\mathcal{EL}^{lin}$ TBoxes, the reduction step generates queries that are never processed by the rewriting step. So, in our case, the reduction step is of no use. However, we keep it in the rewriting algorithm in order to handle future extensions to the ontology language.

Lemma 2. *Let $\mathcal{T}$ be an $\mathcal{EL}^{lin}$ TBox and q a CQ of the form $q(x) \leftarrow A(x, y)$, with $A \in A$. If $q_{rew} \in Rewrite(q, \mathcal{T})$, then q_{rew} is an SPCQ.*

Proof (Sketch). Following from Lemma 1 we know that queries produced by the reduction step are never processed by the rewriting step, and so they are never marked with '1'. So, queries produced by the reduction step are never in the output set, and thus we can ignore the reduction step. The proof is then by induction on the set of queries that are produced after each rewriting step. We denote by $Q^{[i]}$ the set of the queries produced after the i-th iteration of the repeat loop in Algorithm 1.

BASE STEP. $Q^{[1]} = \{q(x) \leftarrow A(x, y)\}$ plus the queries obtained by the 1st rewriting step. The possible cases are the rewriting rules (a), (b) and (c) which generate SPCQs.

INDUCTIVE STEP. If $q \in Q^{[i+1]}$, then q is computed by applying a rewriting rule to a query in $Q^{[i]}$. The claim follows by induction if for each rewriting q to q', we have that q' is a SPCQ. If q is a SPCQ then we can identify a fixed set of possible rewriting cases, according to the rewriting rules. For each possible rewriting case, when q is rewritten to q', it is easy to see that if q is a SPCQ, then q' is a SPCQ.

It is important to note that this algorithm is not guaranteed to terminate on $\mathcal{EL}^{\ell in}$ knowledge bases; see e.g. the TBox of Example 3. This is because the algorithm essentially enumerates all rewritings produced by single applications of the reduction and the rewriting steps; when the algorithm is forced by the TBox to cycle on a set of assertions, it produces infinite branches and does not terminate. However, it is possible to capture such cyclic applications of the rewriting steps if we adopt a more expressive target language for the rewriting; this is the subject of the next two sections: in Sect. 4 we show how to encode the rewritings for an atomic query by means of a finite-state automaton; in Sect. 5 we present a technique, similar to that shown in [14] for OWL QL, that allows us to combine rewritings for atomic queries so as to obtain a CRPQ rewriting for CQs.

4   Rewriting for Atomic Concept Queries

In this section we show how to encode rewritings for atomic queries under $\mathcal{EL}^{\ell in}$ by means of a finite-state automaton; intuitively, the automaton is able to encode infinite sequences of rewriting steps executed according to the algorithm of Sect. 3. We concentrate on atomic queries having a concept atom in the body, since in the case of role atoms, this is done by a simple check on sequences of role inclusions in the TBox — see [13].

Definition 2. *Let $\mathcal{T}$ be an $\mathcal{EL}^{\ell in}$ TBox in normal form, Σ the alphabet $R \cup A$ and A a concept name appearing in $\mathcal{T}$. The NFA-rewriting of A with respect to $\mathcal{T}$, denoted by $\mathsf{NFA}_{A,\mathcal{T}}$, is the NFA over Σ of the form $(Q, \Sigma, \delta, q_0, F)$ defined as follows:*

(1) states S_A and SF_A are in Q, SF_A is in F, and transition (S_A, A, SF_A) is in δ;

(2) S_A is the initial state q_0, $S_\top$ is a final state;

(3) for each $B \in A$ that appears in at least one concept or role inclusion axiom of $\mathcal{T}$, states S_B and SF_B are in Q, SF_B is in F, and transition (S_B, B, SF_B) is in δ;

(4) for each concept inclusion axiom $\rho \in \mathcal{T}$: (4.1) if ρ is of the form $B \sqsubseteq C$, where $B, C \in A$, the transition (S_C, ϵ, S_B) is in δ; (4.2) if ρ is of the form $B \sqsubseteq \exists R.\top$, where $B \in A$ and $R \in R$, for each transition $(S_X, R, S_\top) \in \delta$, the transition (S_X, ϵ, S_B) is in δ; (4.3) if ρ is of the form $\exists R.\top \sqsubseteq B$, where $B \in A$ and $R \in R$, the transition $(S_B, R, S_\top)$ is in δ; (4.4) if ρ is the form $\exists R.D \sqsubseteq C$, where $C, D \in A$ and $R \in R$, the transition (S_C, R, S_D) is in δ;

(4.5) if ρ is the form $C \sqsubseteq \exists R.D$, where $C, D \in A$ and $R \in R$, for any sequence of transitions starting from S_X that accepts the strings RD or R, the transition (S_X, ϵ, S_C) is in δ;

(5) for each role inclusion axiom $T \sqsubseteq S \in \mathcal{T}$ and each transition of the form $(S_C, S, S_B) \in \delta$, the transition (S_C, T, S_B) is in δ.

Example 5. Consider the TBox $\mathcal{T}$ defined by the following inclusion assertions: $\exists R.C \sqsubseteq \exists P.\top$, $\exists P.\top \sqsubseteq A$, $\exists P.\top \sqsubseteq B$, $\exists T.B \sqsubseteq C$ and $\exists S.A \sqsubseteq A$, where P, R, S, T are role names and A, B, C are concept names. Consider now the query $q = q(x) \leftarrow A(x, y)$. First, we transform $\mathcal{T}$ into normal form, say $\mathcal{T}'$, by adding a fresh concept name X and by replacing $\exists R.C \sqsubseteq \exists P.\top$ by $\exists R.C \sqsubseteq X$ and $X \sqsubseteq \exists P.\top$. It is easy to see that $\mathsf{Rewrite}(q, \mathcal{T}')$ runs indefinitely (for instance, we have an infinite loop when rule *(c)* is applied to the atom $A(x, y)$). Let us consider the NFA rewriting of A with respect to $\mathcal{T}'$. We construct $\mathsf{NFA}_{A,\mathcal{T}'}$ as follows: by *(3)* we have the transitions $(S_A, A, SF_A), (S_B, B, SF_B), (S_C, C, SF_C)$ and (S_X, X, SF_X); by *(4.3)* and the inclusion assertions $\exists P.\top \sqsubseteq A$ and $\exists P.\top \sqsubseteq B$, we have the transitions $(S_A, P, S_\top)$ and $(S_B, P, S_\top)$; by *(4.2)* and the inclusion assertion $X \sqsubseteq \exists P.\top$, we have the transitions (S_A, ϵ, S_X) and (S_B, ϵ, S_X); finally, by *(4.4)* and the inclusion assertions $\exists R.C \sqsubseteq X$, $\exists T.B \sqsubseteq C$ and $\exists S.A \sqsubseteq A$, we have the transitions (S_X, R, S_C), (S_C, T, S_B) and (S_A, S, S_A). The NFA $\mathsf{NFA}_{A,\mathcal{T}'}$ is illustrated in Fig. 1. The language accepted by $\mathsf{NFA}_{A,\mathcal{T}'}$ can be described by the following regular expression: $(\exists S.)^*(A|X|(RT)^*(P|RC|RT(B|X)))$. It is easy to see that all the infinite outputs of $\mathsf{Rewrite}(q, \mathcal{T}')$ are of the form $q(x) \leftarrow \mathsf{NFA}_{A,\mathcal{T}'}(x, y)$. For instance, some possible rewritings of q are:

$$q(x) \leftarrow S(x, z_1), S(z_1, z_2), P(z_2, y)$$
$$q(x) \leftarrow S(x, z_1), S(z_1, z_2), A(z_2, y)$$
$$q(x) \leftarrow R(x, z_1), T(z_1, z_2), R(z_2, z_3), C(z_3, y)$$

It is easy to verify that each of these output queries is a SPCQ and each path is in $L(\mathsf{NFA}_{A,\mathcal{T}'})$.

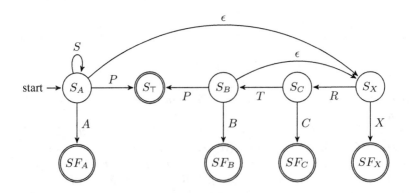

Fig. 1. NFA for Example 5.

Theorem 1. *Let $\mathcal{T}$ be an $\mathcal{EL}^{\ell in}$ TBox, and a concept A. We have that $q \in$ Rewrite$(q(x) \leftarrow A(x,y), \mathcal{T})$ if and only if path$(q) \in L(NFA_{A,\mathcal{T}})$.*

Proof (Sketch). ($\Rightarrow$) The proof is by induction on the set of queries that are marked with '1' after each rewriting step, as the queries marked with '0' are not returned by the algorithm. We denote by $Q^{[i]}$ the set of queries marked with '1' after the i-th application of the rewriting step.

BASE STEP. $Q^{[0]} = \{q\}$. By (2) we have that S_A is the initial state q_0 and by (3) we have the transition (S_A, A, SF_A) therefore $A \in L(NFA_{A,\mathcal{T}})$ and the claim follows trivially.

INDUCTIVE STEP. From Lemma 1 we have that, if an axiom $\rho \in \mathcal{T}$ is not applicable to $body(q)$, for each set of atoms $S \subseteq body(q)$ that unify, ρ is also not applicable to $body(\text{Reduce}(q, S))$. It follows that if a query q' is marked with '0' then there is no axiom in $\mathcal{T}$ that is applicable to q'. Thus, if $q \in Q^{[i+1]}$, then q is computed by applying a rewriting rule to a query that is marked with '1' at i-th application of the rewriting step, which is a query in $Q^{[i]}$. Suppose that for each $q \in Q^{[i]}$ we have that $\text{Path}(q) \in L(NFA_{A,\mathcal{T}})$, the claim follows by induction if for each rewriting q to q', we have that $\text{Path}(q') \in L(NFA_{A,\mathcal{T}})$. From Lemma 2 it follows that the body of each query marked with '1' is a simple path, thus we can identify a fixed set of possible rewriting cases. For each of possible rewriting case, when q is rewritten to q', there is a rule in the definition of $L(NFA_{A,\mathcal{T}})$ such that $\text{Path}(q) \in L(NFA_{A,\mathcal{T}}) \rightarrow \text{Path}(q') \in L(NFA_{A,\mathcal{T}})$ is true.

($\Leftarrow$) The claim follows by induction on the construction rules of the $NFA_{A,\mathcal{T}}$ starting from A, which correspond to all the possible rewriting steps of Rewrite$(q(x) \leftarrow A(x,y), \mathcal{T})$.

Theorem 2. *Given an $\mathcal{EL}^{\ell in}$ TBox $\mathcal{T}$, concept A and a complex concept B, we have that $\mathcal{T} \models B \sqsubseteq A$ if and only if $B \sqsubseteq NFA_{A,\mathcal{T}}$.*

Proof. From Theorem 1 we have that $NFA_{A,\mathcal{T}}$ is a perfect rewriting of A with respect to $\mathcal{T}$ and the claim follows.

5   CRPQ Rewriting for $\mathcal{EL}^{\ell in}$

In this section, following the approach of [13,14], we split the problem of rewriting CQs under $\mathcal{EL}^{\ell in}$ in two: we deal separately with the part of the TBox that does not have existential quantification on the right-hand side of assertions (that is, the part that when expanded does not produce any labelled null) and with the rest of the TBox. We make use of the algorithm for atomic queries presented in the previous section. Then, we put together the solutions devised for the two parts to produce a rewriting algorithm for CQs under $\mathcal{EL}^{\ell in}$.

Given an $\mathcal{EL}^{\ell in}$ knowledge base $(\mathcal{T}, \mathcal{A})$ with $\mathcal{T}$ in normal form, we can find all answers to a CQ q over this KB by evaluating q over the (possibly infinite) canonical model $\mathcal{C}_{\mathcal{T},\mathcal{A}}$ which can be constructed using the chase procedure. We begin by defining the *standard model* $\mathcal{I}_{\mathcal{A}}$ of the ABox $\mathcal{A}$ as follows: (1) $\Delta^{\mathcal{I}_{\mathcal{A}}} = \text{ind}(\mathcal{A})$; (2) $a^{\mathcal{I}_{\mathcal{A}}} = a$, for $a \in \text{ind}(\mathcal{A})$; (3) $A^{\mathcal{I}_{\mathcal{A}}} = \{a \mid A(a) \in \mathcal{A}\}$, for concept name A; *(4)* $P^{\mathcal{I}_\mathcal{A}} = \{(a,b) \mid P(a,b) \in \mathcal{A}\}$, for role name P. Then we take the standard model $\mathcal{I}_\mathcal{A}$ as $\mathcal{I}_0$ and apply inductively the following rules to obtain $\mathcal{I}_{k+1}$ from $\mathcal{I}_k$: *(a')* if $d \in A_1^{\mathcal{I}_k}$ and $A_1 \sqsubseteq A_2 \in \mathcal{T}$, then we add d to $A_2^{\mathcal{I}_{k+1}}$; *(b')* if $(d,d') \in R_1^{\mathcal{I}_k}$ and $R_1 \sqsubseteq R_2 \in \mathcal{T}$, then we add (d,d') to $R_2^{\mathcal{I}_{k+1}}$; *(c')* if $d \in (R.D)^{\mathcal{I}_k}$ and $\exists R.D \sqsubseteq A \in \mathcal{T}$, where D is a concept name or $\top$, then we add d to $A^{\mathcal{I}_{k+1}}$; *(d')* if $d \in A^{\mathcal{I}_k}$ and $A \sqsubseteq \exists R.D \in \mathcal{T}$, where D is a concept name or $\top$, then we take a *fresh* labelled null, d', and add d' to $D^{\mathcal{I}_{k+1}}$ and (d,d') to $R^{\mathcal{I}_{k+1}}$. The canonical model $\mathcal{C}_{\mathcal{T},\mathcal{A}}$ constructed using rules (a'), (b'), (c') and (d') in a bottom-up fashion can alternatively be defined with the top-down approach illustrated in this section; this will be required for query rewriting in Sect. 5.2. There are two key observations that lead us to the alternative definition: first, fresh labelled nulls can only be added by applying (d'), and, second, if two labelled nulls, d_1 and d_2, are introduced by applying (d') with the same concept inclusion $A \sqsubseteq \exists R.D$, then the same rules will be applicable to d_1 and d_2 in the continuation of the chase procedure. So, each labelled null d' resulting from applying (d') to some $A \sqsubseteq \exists R.D$ on a domain element d can be identified with a pair of the form $(d, \exists R.D)$. Following from Theorem 2, for each concept $\exists R.D$ that appears at the RHS of a concept inclusion axiom in $\mathcal{T}$, we introduce a fresh symbol $w_{\exists R.D}$ that is a *witness* for $\exists R.D$ and define a generating relation $\leadsto_{\mathcal{T},\mathcal{A}}$ on the set of these witnesses together with $\mathrm{ind}(\mathcal{A})$ by taking:

- $a \leadsto_{\mathcal{T},\mathcal{A}} w_{\exists R.D}$, if $a \in \mathrm{ind}(\mathcal{A})$, $\mathcal{I}_\mathcal{A} \models B(a)$ and $B \sqsubseteq \mathsf{NFA}_{\mathcal{A},\mathcal{T}}$,
- $w_{\exists S.B} \leadsto_{\mathcal{T},\mathcal{A}} w_{\exists R.D}$ if $B \sqsubseteq \mathsf{NFA}_{\mathcal{A},\mathcal{T}}$ and $A \sqsubseteq \exists R.D \in \mathcal{T}$,

where S is a role name. We point out that we are able to define a finite generating relation $\leadsto_{\mathcal{T},\mathcal{A}}$ for an $\mathcal{EL}^{lin}$ knowledge base $(\mathcal{T},\mathcal{A})$ with the definition of $\mathsf{NFA}_{\mathcal{A},\mathcal{T}}$. In fact, $\mathsf{NFA}_{\mathcal{A},\mathcal{T}}$ captures all (possibly infinite) expressions B such that $\mathcal{T} \models B \sqsubseteq A$. This allows us to exploit the Tree-Witness rewriting technique in [14] (see Sect. 5.2).

A *path*$_{\leadsto_{\mathcal{T},\mathcal{A}}}$ σ is a finite sequence $aw_{\exists R_1.D_1} \cdots w_{\exists R_n.D_n}$, $n \geqslant 0$, such that $a \in \mathrm{ind}(\mathcal{A})$ and, if $n > 0$, then $a \leadsto_{\mathcal{T},\mathcal{A}} w_{\exists R_1.D_1}$ and $w_{\exists R_i.D_i} \leadsto_{\mathcal{T},\mathcal{A}} w_{\exists R_{i+1}.D_{i+1}}$, for $i < n$. Thus, a path of the form $\sigma w_{\exists R.D}$ is also the fresh labelled null introduced by applying (d') to some $A \sqsubseteq \exists R.D$ on the domain element σ (and which corresponds to the pair $(\sigma, \exists R.D)$ mentioned above). Let us denote by $\mathrm{tail}(\sigma)$ the last element in σ; as we noted above, the last element in σ uniquely determines all the subsequent rule applications. The *canonical model* $\mathcal{C}_{\mathcal{T},\mathcal{A}}$ is defined by taking $\Delta^{\mathcal{C}_{\mathcal{T},\mathcal{A}}}$ to be the set of all *path*$_{\leadsto_{\mathcal{T},\mathcal{A}}}$ and taking: *(1)* $a^{\mathcal{C}_{\mathcal{T},\mathcal{A}}} = a$, for $a \in \mathrm{ind}(\mathcal{A})$; *(2)* $A^{\mathcal{C}_{\mathcal{T},\mathcal{A}}} = \{a \in \mathrm{ind}(A) \mid \mathcal{I}_\mathcal{A} \models B(a)$ and $B \sqsubseteq \mathsf{NFA}_{\mathcal{A},\mathcal{T}}\} \cup \{\sigma w_{\exists R.D} \mid D \sqsubseteq \mathsf{NFA}_{\mathcal{A},\mathcal{T}}\}$, for each concept name A; *(3)* $P^{\mathcal{C}_{\mathcal{T},\mathcal{A}}} = \{(a,b) \mid \mathcal{I}_\mathcal{A} \models R(a,b)$ and $\mathcal{T} \models R \sqsubseteq P\} \cup \{(\sigma, \sigma w_{\exists R.D}) \mid \mathrm{tail}(\sigma) \leadsto_{\mathcal{T},\mathcal{A}} w_{\exists R.D}, \mathcal{T} \models R \sqsubseteq P\}$, for a role name P. We point out that, by the definition of rule (b'), we have that $\mathcal{T} \models R \sqsubseteq P$ only if there is a sequence of roles $R_0, \ldots, R_n$ such that $R_{i-1} \sqsubseteq R_i$ are in $\mathcal{T}$, for $1 \leqslant i \leqslant n$, and $R_n = P$. For proof refer to [13].

Given a CQ q, we use the assertions of the TBox $\mathcal{T}$ to rewrite q into another query q' that returns, when evaluated over the data instance (ABox) $\mathcal{A}$, all the certain answers of q with respect to $(\mathcal{T},\mathcal{A})$. Notice that the rewriting q' only depends on the TBox $\mathcal{T}$ and the given query q; it is independent of the ABox $\mathcal{A}$. In query processing, therefore, we use $\mathcal{A}$ only in the final step, when the rewriting is evaluated on it.

We call a CQ q and a TBox $\mathcal{T}$ *CRPQ-rewritable* if there exists a CRPQ q' such that, for any ABox $\mathcal{A}$ and any tuple $\mathbf{a}$ of individuals in $\mathsf{ind}(\mathcal{A})$, we have $(\mathcal{T}, \mathcal{A}) \models q(\mathbf{a})$ if and only if $\mathcal{A} \models q'(\mathbf{a})$.

5.1   Rewriting for Flat $\mathcal{EL}^{\ell in}$

We first consider an important special case of *flat $\mathcal{EL}^{\ell in}$* TBoxes that do not contain existential quantifiers on the right-hand side of concept inclusions. In other words, flat $\mathcal{EL}^{\ell in}$ in normal form can only contain concept and role inclusions of the form $A_1 \sqsubseteq A_2$, $\exists R.D \sqsubseteq A$ and $R_1 \sqsubseteq R_2$, for concept names A, A_1, A_2, role names R_1, R_2, and D a concept name or $\top$. Now let $\mathcal{T}$ be a flat $\mathcal{EL}^{\ell in}$ TBox, q a conjunctive query and a a tuple of individuals. Since $\mathcal{C}_{\mathcal{T},\mathcal{A}}$ is the canonical model for $(\mathcal{T}, \mathcal{A})$, we have that $(\mathcal{T}, \mathcal{A}) \models q(a)$ if and only if $q(a)$ is true in the canonical model $\mathcal{C}_{\mathcal{T},\mathcal{A}}$. The TBox is flat, the generating relation $\rightsquigarrow_{\mathcal{T},\mathcal{A}}$ is empty, the canonical model $\mathcal{C}_{\mathcal{T},\mathcal{A}}$ contains no labelled nulls, and so, by the definition of $\mathcal{C}_{\mathcal{T},\mathcal{A}}$, we have that:

- $\mathcal{C}_{\mathcal{T},\mathcal{A}} \models A(a)$ if and only if $\mathcal{I}_{\mathcal{A}} \models B(a)$ and $\mathcal{T} \models B \sqsubseteq \mathsf{NFA}_{A,\mathcal{T}}$, for some B,
- $\mathcal{C}_{\mathcal{T},\mathcal{A}} \models P(a,b)$ if and only if $\mathcal{I}_{\mathcal{A}} \models R(a,b)$ and $\mathcal{T} \models R \sqsubseteq P$, for some R.

For a CQ q, we define now rewriting q_{ext} as a union of CRPQs which is the result of replacing every atom $A(z_1, z_2)$ in q with $A_{ext}(z_1, z_2)$ and every atom $P(z_1, z_2)$ in q with $P_{ext}(z_1, z_2)$, where $A_{ext}(u_1, u_2) = \mathsf{NFA}_{A,\mathcal{T}}(u_1, u_2)$ and $P_{ext}(u_1, u_2) = \bigcup_{\mathcal{T} \models R \sqsubseteq P} R(u_1, u_2)$. This leads to the following results.

Proposition 1. *For all concept names A, role names P and individual names a and b we have: (1) $\mathcal{C}_{\mathcal{T},\mathcal{A}} \models A(a)$ if and only if $\mathcal{I}_{\mathcal{A}} \models q() \leftarrow A_{ext}(a, a)$, (2) $\mathcal{C}_{\mathcal{T},\mathcal{A}} \models P(a,b)$ if and only if $\mathcal{I}_{\mathcal{A}} \models q() \leftarrow P_{ext}(a, b)$.*

Proof. Follows immediately from the definitions of the formulas A_{ext} and P_{ext}.

Proposition 2. *For any CQ q and any flat $\mathcal{EL}^{\ell in}$ TBox $\mathcal{T}$, q_{ext} is the CRPQ rewriting of q with respect to $\mathcal{T}$.*

Proof. Follows immediately from the previous proposition and from the fact that each formula of the form $R_1(u_1, u_2) \cup \cdots \cup R_n(u_1, u_2)$ can be expressed as a regular path formula of the form $R_1 | \cdots | R_n(u_1, u_2)$.

5.2   Tree-Witness Rewriting for Full $\mathcal{EL}^{\ell in}$

Following a divide and conquer strategy, we show how the process of constructing FO-rewritings can be split into two steps: the first step considers only the flat part of the TBox and uses the formulas $A_{ext}(u_1, u_2)$ and $P_{ext}(u_1, u_2)$ defined in Sect. 5.1; the second step (to be described below) takes account of the remaining part of the TBox, that is, inclusions of the form $A \sqsubseteq \exists R.D$. We first need some preliminary definitions.

Definition 3. *(H-completeness) Let T be a (not necessarily flat) $\mathcal{EL}^{\ell in}$ TBox. A simple ABox $\mathcal{A}$ is said to be H-complete with respect to T if, for all concept names A and role names P, we have:*

- *$A(a) \in \mathcal{A}$ if $\mathcal{I}_{\mathcal{A}} \models B(a)$ and $B \sqsubseteq \mathsf{NFA}_{A,T}$, for some B,*
- *$P(a,b) \in \mathcal{A}$ if $\mathcal{I}_{\mathcal{A}} \models R(a,b)$ and $T \models R \sqsubseteq P$, for some R.*

Observe that, if an ABox $\mathcal{A}$ is H-complete with respect to T, then the ABox part of $\mathcal{C}_{T,\mathcal{A}}$ coincides with $\mathcal{I}_{\mathcal{A}}$. Thus, if T is flat then q itself is clearly the perfect rewriting of q and T over H-complete ABoxes. This leads to the following proposition.

Proposition 3. *If q' is the perfect rewriting of q and T over H-complete ABoxes, then q'_{ext} is the perfect rewriting of q with respect to T.*

So, to generate a CRPQ rewriting we can now focus on constructing rewritings over H-complete ABoxes. To achieve this, we reuse a technique adopted in [14] called *Tree Witness*. Suppose T is a $\mathcal{EL}^{\ell in}$ TBox in normal form. To compute certain answers to q over $(T, \mathcal{A})$, for some $\mathcal{A}$, it is enough to find answers to q in the canonical model $\mathcal{C}_{T,\mathcal{A}}$. To do this, we have to check, for every tuple a of elements in $\mathsf{ind}(\mathcal{A})$, whether there exists a homomorphism from $q(a)$ to $\mathcal{C}_{T,\mathcal{A}}$. Thus, as in the case of flat TBoxes, the answer variables take values from $\mathsf{ind}(A)$. However, the existentially quantified variables in q can be mapped both to $\mathsf{ind}(\mathcal{A})$ and to the labelled nulls in $\mathcal{C}_{T,\mathcal{A}}$. In order to identify how the existential variables can be mapped to the anonymous part, it is sufficient to take a look at the tree-like structure of the generating relation. This technique allows us to rewrite a CQs with respect to $\mathcal{EL}^{\ell in}$ TBoxes over H-complete ABoxes; for details on the Tree Witness rewriting technique, consult [13, 14].

Theorem 3. *Let T be an $\mathcal{EL}^{\ell in}$ TBox and q a CQ. Let q' be the Tree Witness rewriting for q with respect to T over H-complete ABoxes. For any ABox $\mathcal{A}$ and any tuple a in $\mathsf{ind}(\mathcal{A})$, we have $\mathcal{C}_{T,\mathcal{A}} \models q(a)$ if and only if $\mathcal{I}_{\mathcal{A}} \models q'_{ext}(a)$.*

Corollary 1. *Let T be an $\mathcal{EL}^{\ell in}$ TBox and $\mathbf{q}$ a CQ. T is CRPQ-rewritable with respect to $\mathbf{q}$.*

6   Discussion

In this paper we presented a rewriting algorithm for answering conjunctive queries under $\mathcal{EL}^{\ell in}$ knowledge bases. We showed how to encode rewritings of atomic queries with finite-state automata, and finally how to combine such automata in order to produce rewritings for the full language of CQs. We believe that our contribution sheds light on the possibilities of efficient query rewriting under DLs. Our rewriting technique achieves optimal data complexity (see below), and produces a "compact" rewriting without having to take the ABox into account; then the rewriting is evaluated on the ABox, which does not need to be expanded. We therefore argue that this pure rewriting approach is likely to be suitable for real-world cases, especially considering that the final CRPQ evaluation step can be performed by expressing the CRPQ in SPARQL 1.1.

Complexity. When considering query answering under ontologies, the most important asymptotic complexity measure is the so-called *data complexity*, i.e. the complexity w.r.t. the ABox $\mathcal{A}$. Our rewriting is evaluated on $\mathcal{A}$ in NLOGSPACE in data complexity [4], which coincides with the lower bound for CQ answering in $\mathcal{EL}^{\ell in}$ (see [17], where $\mathcal{EL}^{\ell in}$ is called DL-lite$^+$). In terms of *combined complexity*, i.e. the complexity w.r.t. $\mathcal{A}$, $\mathcal{T}$ and q, we limit ourselves to a few consideration, leaving the issue to another work. Our rewriting, which is exponential in the query and the TBox, similarly to other approaches, has the advantage of being expressed in CRPQs (which implies the possibility of being easily translated in SPARQL, as above noted) whose evaluation is in NP [8]. Moreover our technique behaves as "pay-as-you-go" because in the case of atomic queries it produces a rewriting as an RPQ which can be evaluated in NLOGSPACE.

Related Work. Query rewriting has been extensively employed in query answering under ontologies [10,15,16]. In particular, [17] presents a resolution-based query rewriting algorithm for DL-Lite+ ontologies (which is $\mathcal{EL}^{\ell in}$), with Linear Datalog as target language. In [3] the authors introduce a backward chaining mechanism to identify decidable classes of tuple-generating dependencies. Tractable query rewriting (in NLOGSPACE) for the DL-Lite family was presented in [9]; similarly, Rosati [18] used a rewriting algorithm for DL TBoxes expressed in the $\mathcal{EL}$ family of languages [1] to show that query answering in $\mathcal{EL}$ is PTIME-complete in data complexity. Other works [5,6] study FO-rewritability of conjunctive queries in the presence of ontologies formulated in a description logic between $\mathcal{EL}$ and Horn-$\mathcal{SHIF}$, along with related query containment problems. In [11,12] the authors propose an algorithm for computing FO rewritings of concept queries under $\mathcal{EL}$ TBoxes that is tailored towards efficient implementation. The tree-witness technique adopted in this paper is derived from that of [13,14], which address query rewriting over $\mathcal{EL}$, $\mathcal{QL}$ and $\mathcal{RL}$, and propose the tree-witness approach to rewrite $\mathcal{QL}$. The complexity of answering CRPQs under DL-Lite and $\mathcal{EL}$ families is studied in [7].

Future Work. We are extending our work in several directions. The most immediate ones are the following: *(1)* we plan to consider CRPQs as the language for queries, and devise a suitable rewriting algotithm; *(2)* we plan to consider inverse roles, and identify syntactic properties that would still guarantee rewritability of CQs into CRPQs; note that the ontology language so defined subsumes $\mathcal{QL}$; *(3)* we intend to include complex role chains and unions in the language, as in [15]. We already have results on *(1)* and *(3)*.

Acknowledgments. We thank Michael Zakharyaschev and Roman Kontchakov for precious discussions about this material.

References

1. Baader, F., Brandt, S., Lutz, C.: Pushing the EL envelope. In: IJCAI (2005)
2. Baader, F., Nutt, W.: Basic description logics. In: Description Logic Handbook. pp. 43–95 (2003)
3. Baget, J.F., Leclre, M., Mugnier, M.L., Salvat, E.: On rules with existential variables: walking the decidability line. Artif. Intell. **175**(9), 1620–1654 (2011)
4. Barcelo, P., Libkin, L., Lin, A.W., Wood, P.T.: Expressive languages for path queries over graph-structured data. TODS **37**(4), 31 (2012)
5. Bienvenu, M., Hansen, P., Lutz, C., Wolter, F.: First order-rewritability and containment of conjunctive queries in horn description logics. In: DLOG (2016)
6. Bienvenu, M., Lutz, C., Wolter, F.: First-order rewritability of atomic queries in horn description logics. In: IJCAI (2013)
7. Bienvenu, M., Ortiz, M., Simkus, M.: Conjunctive regular path queries in lightweight description logics. In: IJCAI (2013)
8. Bourhis, P., Krötzsch, M., Rudolph, S.: Reasonable highly expressive query languages. In: IJCAI (2015)
9. Calvanese, D., De Giacomo, G., Lembo, D., Lenzerini, M., Rosati, R.: Tractable reasoning and efficient query answering in description logics: The DL-lite family. J. Autom. Reason. **39**(3), 385–429 (2007)
10. Gottlob, G., Orsi, G., Pieris, A.: ontological queries: rewriting and optimization. In: ICDE (2011)
11. Hansen, P., Lutz, C., Seylan, I., Wolter, F.: Query rewriting under EL TBoxes: efficient algorithms. In: Description Logics (2014)
12. Hansen, P., Lutz, C., Seylan, I., Wolter, F.: Efficient query rewriting in the description logic EL and beyond. In: IJCAI (2015)
13. Kikot, S., Kontchakov, R., Zakharyaschev, M.: Conjunctive query answering with OWL 2 QL. In: KR (2012)
14. Kontchakov, R., Zakharyaschev, M.: An introduction to description logics and query rewriting. In: Koubarakis, M., Stamou, G., Stoilos, G., Horrocks, I., Kolaitis, P., Lausen, G., Weikum, G. (eds.) Reasoning Web. LNCS, vol. 8714, pp. 195–244. Springer, Heidelberg (2014)
15. Mosurovic, M., Krdzavac, N., Graves, H., Zakharyaschev, M.: A decidable extension of SROIQ with complex role chains and unions. JAIR **47**, 809–851 (2013)
16. Pérez-Urbina, H., Horrocks, I., Motik, B.: Efficient query answering for OWL 2. In: Bernstein, A., Karger, D.R., Heath, T., Feigenbaum, L., Maynard, D., Motta, E., Thirunarayan, K. (eds.) ISWC 2009. LNCS, vol. 5823, pp. 489–504. Springer, Heidelberg (2009)
17. Schewe, K.-D., Thalheim, B.: Semantics in data and knowledge bases. In: Schewe, K.-D., Thalheim, B. (eds.) SDKB 2008. LNCS, vol. 4925, pp. 1–25. Springer, Heidelberg (2008)
18. Rosati, R.: On conjunctive query answering in EL. In: DL (2007)

Scalable Reasoning by Abstraction Beyond DL-Lite

Birte Glimm, Yevgeny Kazakov, and Trung-Kien Tran$^{(\boxtimes)}$

University of Ulm, Ulm, Germany
{birte.glimm,yevgeny.kazakov,trung-kien.tran}@uni-ulm.de

Abstract. Recently, it has been shown that ontologies with large datasets can be efficiently materialized by a so-called abstraction refinement technique. The technique consists of the *abstraction* phase, which partitions individuals into equivalence classes, and the *refinement* phase, which re-partitions individuals based on entailments for the representative individual of each equivalence class. In this paper, we present an *abstraction*-based approach for materialization in DL-Lite, i.e. we show that materialization for DL-Lite does not require the refinement phase. We further show that the approach is sound and complete even when adding disjunctions and nominals to the language. The proposed technique allows not only for faster materialization and classification of the ontologies, but also for efficient consistency checking; a step that is often omitted by practical approaches based on query rewriting. A preliminary empirical evaluation on both real-life and benchmark ontologies demonstrates that the approach can handle ontologies with large datasets efficiently.

1 Introduction

Over many years, Description Logics (DLs) have been very popular languages for knowledge representation and reasoning. Among the various fragments of Description Logics, DL-Lite [1,3] is a family of languages specifically designed for ontology-based data access (OBDA). In this setting, an ontology with background knowledge (a TBox) can be seen as a conceptual view over data repositories (ABoxes), and data can be accessed via query answering services. Common techniques for query answering in DL-Lite are (pure) *rewriting* [3] and *combined* approaches [5,6,13]. In the rewriting approaches, OBDA systems exploit the background knowledge and rewrite the input query so that the rewritten queries are sufficient to retrieve the complete query answer when evaluated over the unmodified data. As the rewritten queries can be very large or complex [12], several optimization techniques have been proposed with the aim of reducing or simplifying the rewritten queries [2,9,16,17]. Combined approaches complement the pure rewriting approaches; they also work for DL fragments that allow for qualified existential quantification. In contrast to pure rewriting, the combined approaches not only rewrite the input query, but also partially or completely expand the data taking the ontology/schema into account. The latter operation is called *data completion* or *ontology materialization*. It plays an important

© Springer International Publishing Switzerland 2016
M. Ortiz and S. Schlobach (Eds.): RR 2016, LNCS 9898, pp. 77–93, 2016.
DOI: 10.1007/978-3-319-45276-0_7

role in the overall performance of the combined approaches, given the fact that the data is often very large in the OBDA applications. In addition, performing ontology materialization only, OBDA systems are already able to provide the complete answers for instance queries. In this paper, we investigate the application of the novel materialization technique via abstraction refinement [7] for DL-Lite ontologies.

The existing abstraction refinement approach consists of two phases: the *abstraction phase* and the *refinement phase*. In the abstraction phase, individuals in the ABox are partitioned into equivalence classes, which are then used to construct a so-called *abstract* ABox. Entailments of the abstract ABox are transformed to entailments for the original ABox, which might result in some individuals no longer belonging to the same equivalence class. Therefore, the previous steps are repeated in the refinement phase, e.g. individuals are re-partitioned, until, eventually, the fixed-point is reached. The approach presented in this paper can be regarded as an enhancement of the existing abstraction refinement approach tailored towards ontologies in DL-Lite and beyond. We make the following contributions:

- We present an abstraction-based approach for materialization for DL-Lite$_{core}^{\mathcal{H}\sqcup}$, an extension of DL-Lite$_{core}$ with role inclusions and disjunctions. The limited form of existential restrictions in DL-Lite enables an efficient way to transform entailments from the abstract ABox to the original ABox. In addition, the presented approach does not require the refinement phase. This allows not only for faster materialization but also for efficient consistency checking of the ontologies. Query answering only makes sense if the ontology is consistent. Therefore, checking consistency is necessary, but this step is often omitted in many query rewriting systems.[1]
- We show that the presented approach is also sound and complete when adding nominals. Moreover, it can be extended to ontology classification, a non-trivial reasoning task in the presence of nominals.
- We evaluate our approach on both real-life and benchmark ontologies. The empirical results demonstrate that the size of the ABoxes can be reduced by orders of magnitude and, as a result, reasoning via abstraction is often much faster than reasoning over the original ontology.

2   Preliminaries

The syntax of DL-Lite$_{core}^{\mathcal{H}\mathcal{O}\sqcup}$ is defined using a vocabulary consisting of countably infinite disjoint sets N_C of *atomic concepts*, N_O of *nominals*, N_R of *atomic roles*, and N_I of *individuals*. A role is either atomic or an *inverse role* r^-, $r \in N_R$. We define the inverse R^- of a role R by $R^- := r^-$ if $R = r$ and $R^- := r$ if $R = r^-$. Complex *concepts* and *axioms* are defined recursively in Table 1. An *ABox* is a finite set of *concept assertions* of the form $A(a)$ and *role assertions* of the form

---

[1] If $\bot$ is allowed in the language, consistency checking can be reduced to querying instances of $\bot$ but it also requires reasoning over the whole data.

Table 1. The syntax and semantics of DL-Lite$_{core}^{\mathcal{HOU}}$

	Syntax	Semantics
Roles:		
Atomic role	R	$R^{\mathcal{I}} \subseteq \Delta^{\mathcal{I}} \times \Delta^{\mathcal{I}}$
Inverse role	R^-	$\{\langle e,d\rangle \mid \langle d,e\rangle \in R^{\mathcal{I}}\}$
Concepts:		
Atomic concept	A	$A^{\mathcal{I}} \subseteq \Delta^{\mathcal{I}}$
Nominal	o	$o^{\mathcal{I}} \subseteq \Delta^{\mathcal{I}}, \|o^{\mathcal{I}}\| = 1$
Top	$\top$	$\Delta^{\mathcal{I}}$
Bottom	$\bot$	$\emptyset$
Negation	$\neg C$	$\Delta^{\mathcal{I}} \backslash C^{\mathcal{I}}$
conjunction	$C \sqcap D$	$C^{\mathcal{I}} \cap D^{\mathcal{I}}$
Disjunction	$C \sqcup D$	$C^{\mathcal{I}} \cup D^{\mathcal{I}}$
Existential restriction	$\exists R$	$\{d \mid \exists e \in \Delta^{\mathcal{I}} : \langle d,e\rangle \in R^{\mathcal{I}}\}$
Axioms:		
Concept inclusion	$C \sqsubseteq D$	$C^{\mathcal{I}} \subseteq D^{\mathcal{I}}$
Role inclusion	$R \sqsubseteq S$	$R^{\mathcal{I}} \subseteq S^{\mathcal{I}}$
Concept assertion	$A(a)$	$a^{\mathcal{I}} \in A^{\mathcal{I}}$
Role assertion	$R(a,b)$	$\langle a^{\mathcal{I}}, b^{\mathcal{I}}\rangle \in R^{\mathcal{I}}$

$R(a,b)$ with $A \in N_C, R \in N_R \cup \{r^- \mid r \in N_R\}$, and $a,b \in N_I$. A *TBox* is a finite set of role and concept inclusions. An *ontology* $\mathcal{O}$, written as $\mathcal{O} = \mathcal{A} \cup \mathcal{T}$, consists of an ABox $\mathcal{A}$ and a TBox $\mathcal{T}$. W.l.o.g. we do not distinguish between the axioms $R(a,b)$ and $R^-(b,a)$ as well as $R \sqsubseteq S$ and $R^- \sqsubseteq S^-$. We use con($\mathcal{O}$), rol($\mathcal{O}$), ind($\mathcal{O}$), nom($\mathcal{O}$) for the sets of atomic concepts, atomic roles, individuals, and nominals occurring in $\mathcal{O}$, respectively. By DL-Lite$_{core}^{\mathcal{HU}}$ we denote the fragment of DL-Lite$_{core}^{\mathcal{HOU}}$ that disallows nominals.

An *interpretation* $\mathcal{I} = (\Delta^{\mathcal{I}}, \cdot^{\mathcal{I}})$ consists of a non-empty set $\Delta^{\mathcal{I}}$, the *domain* of $\mathcal{I}$, and an *interpretation function* $\cdot^{\mathcal{I}}$, that assigns to each $A \in N_C$ a subset $A^{\mathcal{I}} \subseteq \Delta^{\mathcal{I}}$, to each $o \in N_O$ a singleton subset $o^{\mathcal{I}} \subseteq \Delta^{\mathcal{I}}, \|o^{\mathcal{I}}\| = 1$, to each $r \in N_R$ a binary relation $r^{\mathcal{I}} \subseteq \Delta^{\mathcal{I}} \times \Delta^{\mathcal{I}}$, and to each $a \in N_I$ an element $a^{\mathcal{I}} \in \Delta^{\mathcal{I}}$. This assignment is extended to roles and to complex concepts as shown in Table 1. An interpretation $\mathcal{I}$ *satisfies* an axiom α (written $\mathcal{I} \models \alpha$) if the corresponding condition in Table 1 holds. Given an ontology $\mathcal{O}$, $\mathcal{I}$ is a *model* of $\mathcal{O}$ (written $\mathcal{I} \models \mathcal{O}$) if $\mathcal{I} \models \alpha$ for all axioms $\alpha \in \mathcal{O}$; $\mathcal{O}$ *is consistent* if $\mathcal{O}$ has a model; and $\mathcal{O}$ *entails* an axiom α (written $\mathcal{O} \models \alpha$), if every model of $\mathcal{O}$ satisfies α.

For an ontology $\mathcal{O}$, we say that $\mathcal{O}$ is *concept-materialized* if $\mathcal{O} \models A(a)$ implies $A(a) \in \mathcal{O}$ for each $A \in$ con($\mathcal{O}$) and $a \in$ ind($\mathcal{O}$); $\mathcal{O}$ is *role-materialized* if $\mathcal{O} \models r(a,b)$ implies $r(a,b) \in \mathcal{O}$ for each $r \in$ rol($\mathcal{O}$) and $a,b \in$ ind($\mathcal{O}$); $\mathcal{O}$ is (fully) *materialized* if it is both concept and role materialized. The concept-, role-, and/or (full) materialization of an ontology $\mathcal{O}$ is the smallest super-set of

$\mathcal{O}$ that is concept-, role-, and/or fully materialized respectively. Given an ontology, traditional reasoning tasks include *ontology materialization*: computing the materialization of the ontology, *ontology classification*: computing all entailed concept inclusions between atomic concepts in the ontology, and *consistency checking*: checking if the ontology is consistent.

3 Reasoning by Abstraction

The general idea of reasoning via abstraction is to reduce reasoning over a large ABox to reasoning over a smaller one. Specifically, one first builds a suitable *abstraction* of the *original* ontology; performs reasoning over the abstraction; and then *transfers* entailments of the abstraction to corresponding entailments of the original ontology. Correctness of the reduction is based on homomorphisms between ABoxes.

Definition 1. *Let $\mathcal{A}$ and $\mathcal{B}$ be ABoxes. A mapping $h : \mathsf{ind}(\mathcal{B}) \to \mathsf{ind}(\mathcal{A})$ is called a homomorphism (from $\mathcal{B}$ to $\mathcal{A}$) if, for every assertion $\alpha \in \mathcal{B}$, we have $h(\alpha) \in \mathcal{A}$, where $h(C(a)) := C(h(a))$ and $h(R(a,b)) := R(h(a), h(b))$.*

Example 1. Consider the ABoxes $\mathcal{A} = \{A(a),\ A(b),\ R(a,b)\}$, $\mathcal{B}_1 = \{A(u)\}$, and $\mathcal{B}_2 = \{A(v),\ R(v,v)\}$ visualized in Fig. 1. Then the mappings $h_1 = \{u \mapsto a\}$ and $h_2 = \{u \mapsto b\}$ are homomorphisms from $\mathcal{B}_1$ to $\mathcal{A}$; and the mapping $h_3 = \{a \mapsto v,\ b \mapsto v\}$ is a homomorphism from $\mathcal{A}$ to $\mathcal{B}_2$.

The following property of homomorphisms allows us to establish the relation between entailments of one ontology and those of the other.

Lemma 1. *Let $\mathcal{A}$ and $\mathcal{B}$ be ABoxes, and $h : \mathsf{ind}(\mathcal{B}) \to \mathsf{ind}(\mathcal{A})$ a homomorphism from $\mathcal{B}$ to $\mathcal{A}$. Then, for every TBox $\mathcal{T}$ and every axiom α, $\mathcal{B} \cup \mathcal{T} \models \alpha$ implies $\mathcal{A} \cup \mathcal{T} \models h(\alpha)$.*

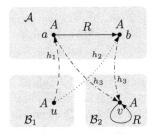
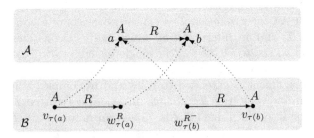

Fig. 1. Visualization of the ABoxes and homomorphisms in Example 1

Fig. 2. Visualization of the ABoxes $\mathcal{A}$ from Example 4 and its abstraction $\mathcal{B}$ from Example 5, where the dotted lines show the homomorphism from $\mathcal{B}$ to $\mathcal{A}$ induced by the abstraction

Note that Lemma 1 is not restricted to DL-Lite$_{core}^{\mathcal{HOU}}$ and it holds for any DL with (classical) set-theoretic semantics, e.g. $\mathcal{SROIQ}$ [10]. The following two corollaries illustrate aspects of homomorphisms that are of particular relevance for our approach, namely that consistency and (concept) entailments are preserved under homomorphisms.

Corollary 1. *Let $\mathcal{A}$ and $\mathcal{B}$ be ABoxes, $h\colon \mathsf{ind}(\mathcal{B}) \to \mathsf{ind}(\mathcal{A})$ a homomorphism from $\mathcal{B}$ to $\mathcal{A}$, $a \in \mathsf{ind}(\mathcal{A})$ and $b \in \mathsf{ind}(\mathcal{B})$ such that $h(b) = a$. Then, for every TBox $\mathcal{T}$ and concept C, $\mathcal{B} \cup \mathcal{T} \models C(b)$ implies $\mathcal{A} \cup \mathcal{T} \models C(a)$.*

Corollary 2. *Let $\mathcal{A}$ and $\mathcal{B}$ be ABoxes. If there exists a homomorphism from $\mathcal{B}$ to $\mathcal{A}$ then, for every TBox $\mathcal{T}$, $\mathcal{A} \cup \mathcal{T}$ is consistent implies $\mathcal{B} \cup \mathcal{T}$ is consistent.*

The abstraction is obtained by partitioning individuals in the original ABox into equivalence classes and by using just one *representative* individual for each equivalence class. Entailments of the representatives are then *transferred* to the corresponding entailments for individuals in the equivalence classes.

If we use the individual u in Example 1 as the representative for a and b, then, for any TBox $\mathcal{T}$, one can *transfer* any newly entailed concept assertion for u to the corresponding assertions for a and b by Corollary 1. However, not all entailments for a and b can necessarily be computed this way.

Example 2 (Example 1 continued). Consider a TBox $\mathcal{T} = \{A \sqsubseteq C, \exists R^- \sqsubseteq B\}$. We have $\mathcal{B}_1 \cup \mathcal{T} \models C(u)$. By Corollary 1, we obtain $C(a)$, $C(b)$ entailed by $\mathcal{A} \cup \mathcal{T}$. We are, however, not able to obtain $B(b)$ via homomorphisms from $\mathcal{B}_1$ to $\mathcal{A}$, although $B(b)$ is entailed by $\mathcal{A} \cup \mathcal{T}$.

Also in Example 1, since there is a homomorphism from $\mathcal{A}$ to $\mathcal{B}_2$, for any TBox $\mathcal{T}$, if $\mathcal{B}_2 \cup \mathcal{T}$ is consistent then $\mathcal{A} \cup \mathcal{T}$ is consistent by Corollary 2. Furthermore, if we use the individual v as the representative for a and b (ignoring that there is no homomorphism from $\mathcal{B}_2$ to $\mathcal{A}$), then we can compute all entailments for a and b based on the entailments of v. However, we might transfer facts that are not entailed by $\mathcal{A} \cup \mathcal{T}$.

Example 3 (Example 2 continued). We have $\mathcal{B}_2 \cup \mathcal{T} \models \{B(v), C(v)\}$. If we take v as representative of both a and b, then we obtain $B(a)$, $B(b)$, $C(a)$, $C(b)$. However, $B(a)$ is not entailed by $\mathcal{A} \cup \mathcal{T}$.

As demonstrated in Examples 2 and 3, it is often easy to obtain either sound or complete results but it is challenging to obtain both. The SHER approach [4] addresses this issue by computing complete but possibly unsound entailments of the ontology using a compressed, so-called summary ABox and by using justification techniques [11] to refine the summary. The abstraction refinement approach [7] computes sound but possibly incomplete entailments. To ensure completeness further refinement steps are employed based on the newly derived entailments. In the next section, we present an enhancement of the existing abstraction refinement approach that is only based on the abstraction. We show that indeed no refinement is needed to obtain both sound and complete entailments for DL-Lite$_{core}^{\mathcal{HOU}}$ ontologies. To simplify presentation, we first present the solution for DL-Lite$_{core}^{\mathcal{HU}}$ and then discuss the extensions for DL-Lite$_{core}^{\mathcal{HOU}}$.

4   Abstraction for DL-Lite$_{core}^{\mathcal{H}\sqcup}$

To construct the abstraction of the original ABox, we partition individuals in the original ABox into equivalence classes and use just one *representative* individual for each equivalence class. The equivalence classes are characterized by the *type* of individuals, which can be syntactically computed from the original ABox.

Definition 2. *Let $\mathcal{A}$ be an ABox and a an individual. The* type *of a (w.r.t. $\mathcal{A}$) is a pair $\tau(a) = \langle \tau_C(a), \tau_R(a)\rangle$ where $\tau_C(a) = \{A \mid A(a) \in \mathcal{A}\}$ and $\tau_R(a) = \{R \mid \exists b : R(a, b) \in \mathcal{A}\}$.*

Example 4. Let $\mathcal{A} = \{A(a), A(b), R(a, b)\}$ be as in Example 1 (cf. Fig. 1). Then, we have $\tau(a) = \langle\{A\}, \{R\}\rangle$ and $\tau(b) = \langle\{A\}, \{R^-\}\rangle$.

The abstract ABox is then constructed by introducing one representative and the respective assertions for each type.

Definition 3. *The* abstraction *of an ABox $\mathcal{A}$ is an ABox $\mathcal{B} = \bigcup_{a \in \mathsf{ind}(\mathcal{A})} \mathcal{B}_{\tau(a)}$, where, for each type $\tau(a) = \langle\tau_C, \tau_R\rangle$, $\mathcal{B}_{\tau(a)} = \{A(v_{\tau(a)}) \mid A \in \tau_C\} \cup \{R(v_{\tau(a)}, w_{\tau(a)}^R) \mid R \in \tau_R\}$, where $v_{\tau(a)}$ and $w_{\tau(a)}^R$ are fresh, distinguished abstract individuals for each type $\tau(a)$.*

Example 5. The abstraction for $\mathcal{A}$ in Example 4 is the ABox $\mathcal{B} = \mathcal{B}_{\tau(a)} \cup \mathcal{B}_{\tau(b)}$, where $\mathcal{B}_{\tau(a)} = \{A(v_{\tau(a)}), R(v_{\tau(a)}, w_{\tau(a)}^R)\}$, $\mathcal{B}_{\tau(b)} = \{A(v_{\tau(b)}), R^-(v_{\tau(b)}, w_{\tau(b)}^{R^-})\}$ (cf. Fig. 2).

Note that the size of the abstraction of a small ABox may be larger than the size of the original ABox, but for ontologies with a large ABox, many individuals have the same type and, hence, abstractions are small.

Intuitively, the abstraction of an ABox is a disjoint union of small ABoxes witnessing each individual type realized in the ABox. There always exist homomorphisms from the abstraction to the original Abox.

Definition 4. *Let $\mathcal{A}$ be an ABox and $\mathcal{B}$ its abstraction as in Definition 3. The abstraction $\mathcal{B}$ induces* a mapping $h : \mathsf{ind}(\mathcal{B}) \to \mathsf{ind}(\mathcal{A})$ *such that:*

$$h(v_\tau) \in \{a \in \mathsf{ind}(\mathcal{A}) \mid \tau(a) = \tau\},$$
$$h(w_\tau^R) \in \{b \in \mathsf{ind}(\mathcal{A}) \mid R(h(v_\tau), b) \in \mathcal{A}\}.$$

Lemma 2. *Let $\mathcal{A}$ be an ABox and $\mathcal{B}$ the abstraction of $\mathcal{A}$. Then, for every mapping h induced by $\mathcal{B}$, h is a homomorphism from $\mathcal{B}$ to $\mathcal{A}$.*

Proof. The mapping h is a homomorphism from $\mathcal{B}$ to $\mathcal{A}$ since, for every $C(v_\tau) \in \mathcal{B}$, we have $h(C(v_\tau)) = C(a) \in \mathcal{A}$ and, for every $R(v_\tau, w_\tau^R) \in \mathcal{B}$, we have $h(R(v_\tau, w_\tau^R)) = R(a, b) \in \mathcal{A}$ for some a, b. $\square$

Once the abstract ABox $\mathcal{B}$ of the original ABox $\mathcal{A}$ has been constructed, instead of performing reasoning over $\mathcal{A}$, we perform reasoning over $\mathcal{B}$ and transfer entailments from the abstraction back to the original ABox using Corollary 1. Intuitively, for each type τ, the abstract individual v_τ is the representative for all individuals of this type. Therefore, for every TBox $\mathcal{T}$ and each $A(v_\tau)$ entailed by $\mathcal{B} \cup \mathcal{T}$, we obtain $A(a)$, where $\tau(a) = \tau$, is entailed by $\mathcal{A} \cup \mathcal{T}$. This gives rise to a procedure for computing the concept materialization of an ontology, which we present in Algorithm 1.

Since materializing an inconsistent ontology would extend the ABox with all possible assertions for the atomic concepts and roles, and individuals used in the ontology, we can furthermore observe that $\mathcal{B} \cup \mathcal{T}$ can also be used to check consistency of $\mathcal{A} \cup \mathcal{T}$. We use this to devise a procedure for checking consistency of an ontology in Algorithm 2. In practice the steps performed by this algorithm can also be directly integrated into Algorithm 1.

Algorithm 1. Procedure for computing the concept materialization of an ontology

---

Input: An ontology $\mathcal{O} = \mathcal{A} \cup \mathcal{T}$
Output: Returns the concept materialized ontology $\mathcal{O}$
1: Compute the abstraction $\mathcal{B}$ of $\mathcal{A}$ according to Definition 3
2: Compute the concept materialization $\mathcal{B}' \cup \mathcal{T}$ of $\mathcal{B} \cup \mathcal{T}$
3: $\Delta\mathcal{B} = \{A(v_\tau) \in \mathcal{B}' \mid A(v_\tau) \notin \mathcal{B}\}$
4: **for all** $A(v_\tau) \in \Delta\mathcal{B}$ **do**
5:     **for all** $a \in \mathrm{ind}(\mathcal{A})$ s.t. $\tau(a) = \tau$ **do**
6:         $\mathcal{A} = \mathcal{A} \cup \{A(a)\}$
7:     **end for**
8: **end for**
9: **return** $\mathcal{A} \cup \mathcal{T}$

Algorithm 2. Procedure for checking consistency of an ontology

---

Input: An ontology $\mathcal{O} = \mathcal{A} \cup \mathcal{T}$
Output: Returns *true* if $\mathcal{O}$ is consistent and *false* otherwise
1: Compute the abstraction $\mathcal{B}$ of $\mathcal{A}$ according to Definition 3
2: **if** $\mathcal{B} \cup \mathcal{T}$ is inconsistent **then**
3:     **return** *false*
4: **else**
5:     **return** *true*
6: **end if**

Soundness of the algorithms follows directly from our previously shown results.

Lemma 3 (Soundness). *Let $\mathcal{A}$ be an ABox, $\mathcal{B}$ its abstraction, and $\mathcal{T}$ a TBox. Then, we have:*

(1) $\mathcal{B} \cup \mathcal{T}$ *is inconsistent implies* $\mathcal{A} \cup \mathcal{T}$ *is inconsistent;*
(2) *for every type* τ *and every concept* C, $\mathcal{B} \cup \mathcal{T} \models C(v_\tau)$ *implies* $\mathcal{A} \cup \mathcal{T} \models C(a)$,
   where $a \in \mathsf{ind}(\mathcal{A})$ *s.t.* $\tau(a) = \tau$.

Proof. The lemma is a straightforward consequence of the fact that the abstraction induces homomorphisms to the original ABox according to Lemma 2. Applying the contrapositive of Corollaries 2 and 1 yields the desired result and, hence, soundness of the algorithms.   □

Example 6. Consider $\mathcal{O} = \mathcal{A} \cup \mathcal{T}$ with $\mathcal{T} = \{A \sqsubseteq C, \exists R^- \sqsubseteq B\}$ from Example 2 and $\mathcal{A}$ from Example 1 as input for Algorithm 1. Figure 2 visualizes $\mathcal{A}$ and its abstraction $\mathcal{B}$. By materializing $\mathcal{B} \cup \mathcal{T}$, we obtain $\Delta\mathcal{B} = \{C(v_{\tau(a)}), C(v_{\tau(b)}), B(v_{\tau(b)})\}$ in Line 3. Note that while $B(w^R_{\tau(a)})$ is in the materialized abstraction $\mathcal{B}'$, it is not part of $\Delta\mathcal{B}$. By updating $\mathcal{A}$ using $\Delta\mathcal{B}$ (Lines 4 to 8), we obtain $\mathcal{A} = \mathcal{A} \cup \{C(a), C(b), B(b)\}$, where all added concept assertions are entailed by the original ontology.

The procedure in Algorithm 1 differs from the abstraction refinement procedure in the existing approach for Horn $\mathcal{ALCHOI}$ [7] in that, for each type τ, only assertions of v_τ are used to update the original ABox. As demonstrated in Example 6, although $B(w^R_{\tau(a)}) \in \mathcal{B}'$, it is not in $\Delta\mathcal{B}$ and, hence, it is not used for extending $\mathcal{A}$. In addition, unlike the algorithm in the existing approach, Algorithm 1 incorporates no refinement step, i.e. there is no repetition of Lines 1–8 until no new assertions can be added to the original ABox $\mathcal{A}$. Such a repetition is required to obtain completeness for the Horn $\mathcal{ALCHOI}$ procedure. We next show that the current procedure is nevertheless *complete* for DL-Lite$^{\mathcal{HU}}_{core}$, that is, the resulting ontology is (concept) materialized when the procedure terminates.

We can immediately show soundness of the algorithms as there always exist homomorphisms from the abstraction $\mathcal{B}$ to the corresponding original ABox $\mathcal{A}$ as in Definition 4. But we do not have a similar property for completeness, i.e. there might exist no homomorphism from $\mathcal{A}$ to $\mathcal{B}$. To show completeness, we construct an extension of $\mathcal{B}$ such that there exists a homomorphism from $\mathcal{A}$ to the extension that maps a to $v_{\tau(a)}$ for each individual $a \in \mathsf{ind}(\mathcal{A})$; and we show that the abstraction entails exactly the same concept assertions as its extension does.

Example 7 (Example 6 continued). Let $\mathcal{B}^+$ be an ABox obtained from $\mathcal{B}$ in Example 6 by adding the role assertion $R(v_{\tau(a)}, v_{\tau(b)})$, and h a mapping from $\mathcal{A}$ to $\mathcal{B}^+$ defined as $h(a) = v_{\tau(a)}, h(b) = v_{\tau(b)}$. Since $h(\mathcal{A}) = \{A(v_{\tau(a)}), A(v_{\tau(b)}), R(v_{\tau(a)}, v_{\tau(b)})\} \subseteq \mathcal{B}^+$, h is a homomorphism from $\mathcal{A}$ to $\mathcal{B}^+$. Therefore, using Corollary 1, we can obtain all entailed assertions of a and b based on entailed assertions of $v_{\tau(a)}$ and $v_{\tau(b)}$ w.r.t. $\mathcal{B}^+ \cup \mathcal{T}$. Furthermore, $\mathcal{B}^+ \cup \mathcal{T}$ and $\mathcal{B} \cup \mathcal{T}$ entail the same set of concept assertions. Hence, the abstraction $\mathcal{B}$ is sufficient for obtaining all entailed assertions of $\mathcal{A} \cup \mathcal{T}$. Indeed, the ABox $\mathcal{A}$ after updating already contains all entailed concept assertions.

As demonstrated in Example 7, for this particular TBox and ABox, the abstraction is sufficient to obtain all entailed concept assertions of the original ontology. In the following lemma, we show that the same property holds for any TBox and ABox.

Lemma 4. *Let $\mathcal{O} = \mathcal{A} \cup \mathcal{T}$ be a DL-Lite$_{core}^{\mathcal{H}\sqcup}$ ontology, $\mathcal{B}$ the abstraction of $\mathcal{A}$, and $\mathcal{B}^+ = \mathcal{B} \cup \{R(v_{\tau(a)}, v_{\tau(b)}) \mid R(a,b) \in \mathcal{A}\}$. Then, we have:*

(1) $\mathcal{B} \cup \mathcal{T}$ is consistent implies $\mathcal{B}^+ \cup \mathcal{T}$ is consistent;

(2) for every atomic concept A and individual v, $\mathcal{B}^+ \cup \mathcal{T} \models A(v)$ implies $\mathcal{B} \cup \mathcal{T} \models A(v)$.

Proof. If $\mathcal{B} \cup \mathcal{T}$ is inconsistent, then the lemma trivially holds. We assume $\mathcal{B} \cup \mathcal{T}$ is consistent and let $\mathcal{I}$ be an arbitrary model of $\mathcal{B} \cup \mathcal{T}$. Next, we construct a model $\mathcal{J}$ of $\mathcal{B}^+ \cup \mathcal{T}$ such that $\mathcal{J} \models A(v)$ implies $\mathcal{I} \models A(v)$ for every atomic concept A and individual v. Then it follows that $\mathcal{B}^+ \cup \mathcal{T}$ is consistent, i.e. Claim (1) holds, and $\mathcal{B}^+ \cup \mathcal{T} \models A(v)$ implies $\mathcal{J} \models A(v)$, which implies $\mathcal{I} \models A(v)$. Since $\mathcal{I}$ is arbitrary, we obtain $\mathcal{B} \cup \mathcal{T} \models A(v)$, i.e. Claim (2) holds. Such a model $\mathcal{J}$ is obtained from $\mathcal{I}$ by setting $\Delta^{\mathcal{J}} = \Delta^{\mathcal{I}}$ and defining the interpretation function as follows:

$$v^{\mathcal{J}} = v^{\mathcal{I}} \text{ for every individual } v$$

$$A^{\mathcal{J}} = A^{\mathcal{I}} \text{ for every atomic concept } A$$

$$r^{\mathcal{J}} = r^{\mathcal{I}} \cup \{\langle v_{\tau(a)}^{\mathcal{I}}, v_{\tau(b)}^{\mathcal{I}} \rangle \mid R(v_{\tau(a)}, v_{\tau(b)}) \in \mathcal{B}^+ \text{ and } \mathcal{O} \models R \sqsubseteq r\}$$

$$\cup \{\langle v_{\tau(b)}^{\mathcal{I}}, v_{\tau(a)}^{\mathcal{I}} \rangle \mid R(v_{\tau(a)}, v_{\tau(b)}) \in \mathcal{B}^+ \text{ and } \mathcal{O} \models R \sqsubseteq r^-\}$$

$$\text{for every atomic role } r$$

We will show $\mathcal{J} \models \mathcal{B}^+ \cup \mathcal{T}$ by showing that it entails every axiom in $\mathcal{B}^+ \cup \mathcal{T}$. Since $\mathcal{I} \models \mathcal{B} \cup \mathcal{T}$ and the interpretation of atomic concepts and individuals remains the same in $\mathcal{J}$, we have $\mathcal{J}$ entails every concept assertion in $\mathcal{B}^+$. And, clearly, from the definition of $\mathcal{J}$, it follows that $\mathcal{J}$ entails every role assertion in $\mathcal{B}^+$.

We now show by induction that, for every DL-Lite$_{core}^{\mathcal{H}\sqcup}$ concept C, we have $C^{\mathcal{J}} = C^{\mathcal{I}}$. Then, for every concept inclusion $C \sqsubseteq D \in \mathcal{T}$, we have $C^{\mathcal{J}} = C^{\mathcal{I}} \subseteq D^{\mathcal{I}} = D^{\mathcal{J}}$, i.e. $\mathcal{J} \models C \sqsubseteq D$.

- Cases $C = A \mid \neg A \mid \top \mid \bot$ are trivial as the interpretation of atomic concepts in $\mathcal{I}$ and in $\mathcal{J}$ are identical.
- Case $C = \exists r$, where $r \in N_R$; the case $\exists r^-$ is symmetric. We have $d \in (\exists r)^{\mathcal{J}}$ iff there exists $e \in \Delta^{\mathcal{J}}$ s.t. $\langle d, e \rangle \in r^{\mathcal{J}}$. If $\langle d, e \rangle \in r^{\mathcal{I}}$, then $d \in (\exists r)^{\mathcal{I}}$. Otherwise, from the definition of $\mathcal{J}$, $\langle d, e \rangle$ results from one of the cases in the role extension. We consider the case $d = v_{\tau(a)}^{\mathcal{I}}, e = v_{\tau(b)}^{\mathcal{I}}$ for some individuals a and b, where $R(v_{\tau(a)}, v_{\tau(b)}) \in \mathcal{B}^+, \mathcal{O} \models R \sqsubseteq r, \mathcal{O} \not\models R \sqsubseteq r^-$; other cases are analogous. By definition of $\mathcal{B}^+$, we have $R(v_{\tau(a)}, v_{\tau(b)}) \in \mathcal{B}^+$ iff $R(a,b) \in \mathcal{A}$. This is the case iff $R(v_{\tau(a)}, w_{\tau(a)}^R) \in \mathcal{B}$ by Definition 3. Since $\mathcal{O} \models R \sqsubseteq r$ and $\mathcal{I} \models \mathcal{B}$, we obtain $\langle v_{\tau(a)}^{\mathcal{I}}, (w_{\tau(a)}^R)^{\mathcal{I}} \rangle \in r^{\mathcal{I}}$, i.e. $d = v_{\tau(a)}^{\mathcal{I}} \in (\exists r)^{\mathcal{I}}$. Since d is arbitrary, we have $(\exists r)^{\mathcal{J}} = (\exists r)^{\mathcal{I}}$.
- Case $C = \neg D$. By induction hypothesis $D^{\mathcal{J}} = D^{\mathcal{I}}$ and since $\Delta^{\mathcal{J}} = \Delta^{\mathcal{I}}$, we have $(\neg D)^{\mathcal{J}} = \Delta^{\mathcal{J}} \setminus D^{\mathcal{J}} = \Delta^{\mathcal{I}} \setminus D^{\mathcal{I}} = (\neg D)^{\mathcal{I}}$, i.e. $C^{\mathcal{J}} = C^{\mathcal{I}}$.
- Cases $C = C_1 \sqcup C_2$ and $C = C_1 \sqcap C_2$. By induction hypothesis, we have $C_1^{\mathcal{J}} = C_1^{\mathcal{I}}$ and $C_2^{\mathcal{J}} = C_2^{\mathcal{I}}$. Therefore, $(C_1 \sqcup C_2)^{\mathcal{J}} = C_1^{\mathcal{J}} \cup C_2^{\mathcal{J}} = C_1^{\mathcal{I}} \cup C_2^{\mathcal{I}} = (C_1 \sqcup C_2)^{\mathcal{I}}$. Similarly, we obtain $(C_1 \sqcap C_2)^{\mathcal{J}} = (C_1 \sqcap C_2)^{\mathcal{I}}$.

For every role inclusion $R \sqsubseteq S \in \mathcal{T}$, by the definition of $\mathcal{J}$ and from $\mathcal{I} \models R \sqsubseteq S$, we have $\mathcal{J} \models R \sqsubseteq S$, which proves $\mathcal{J} \models \mathcal{B}^+ \cup \mathcal{T}$ and, hence, finishes this proof. $\square$

Using Lemma 4, we can establish completeness of Algorithms 1 and 2.

Lemma 5 (Completeness). *Let $\mathcal{A}$ be an ABox, $\mathcal{B}$ its abstraction, and $\mathcal{T}$ a DL-Lite$_{core}^{\mathcal{HU}}$ TBox, then we have:*

(1) $\mathcal{B} \cup \mathcal{T}$ is consistent implies $\mathcal{A} \cup \mathcal{T}$ is consistent;
(2) for every atomic concept A and individual a, $\mathcal{A} \cup \mathcal{T} \models A(a)$ implies $\mathcal{B} \cup \mathcal{T} \models A(v_{\tau(a)})$.

Proof. Let $\mathcal{B}^+$ be the ABox in Lemma 4 and h a mapping from $\mathcal{A}$ to $\mathcal{B}^+$ s.t. $h(a) = v_{\tau(a)}$, for every $a \in \mathsf{ind}(\mathcal{A})$. By the definitions of $\mathcal{B}$ and of $\mathcal{B}^+$, for each $A(a) \in \mathcal{A}$, we have $A(v_{\tau(a)}) \in \mathcal{B}$, which implies $A(v_{\tau(a)}) \in \mathcal{B}^+$. By the definition of $\mathcal{B}^+$, for each $R(a,b) \in \mathcal{A}$, we have $R(v_{\tau(a)}, v_{\tau(b)}) \in \mathcal{B}^+$. Hence, h is a homomorphism from $\mathcal{A}$ to $\mathcal{B}^+$. By Claim (1) of Lemma 4 and Corollary 2, consistency of $\mathcal{B} \cup \mathcal{T}$ implies consistency of $\mathcal{B}^+ \cup \mathcal{T}$, which implies consistency of $\mathcal{A} \cup \mathcal{T}$, i.e. Claim (1) holds. Similarly, by Corollary 1 and Claim (2) of Lemma 4, we have, for each atomic concept A and individual a, $\mathcal{A} \cup \mathcal{T} \models A(a)$ implies $\mathcal{B}^+ \cup \mathcal{T} \models h(A(a))$. Since $h(A(a)) = A(v_{\tau(a)})$, this implies $\mathcal{B} \cup \mathcal{T} \models A(v_{\tau(a)})$ and Claim (2) holds. $\square$

5   Implementation and Evaluation

We have implemented a prototype system Orar[2] for reasoning in DL-Lite$_{core}^{\mathcal{HU}}$. To evaluate the feasibility of our approach, we tested Orar on several real-life and benchmark ontologies and compared the performance of Orar with that of the other popular reasoners. The empirical evaluation results show the approach can reduce the size of the ABoxes significantly (by orders of magnitude), which results in great performance improvements.

The test ontologies are from popular benchmarks and also used in the evaluations of other approaches. NPD[3] is an ontology about petroleum activities, DBPedia$^+$[4] is an extension of the DBPedia ontology, and IMDb[5] consists of the Movie ontology and the dataset extracted from the IMDb website. While NPD, DBPedia$^+$, and IMDb contain real-life data, LUBM and UOBM are popular benchmarks with synthetic data of the university domain. The datasets in LUBM and UOBM can be generated in arbitrary sizes, indicated by the number of universities. We use LUBM n and UOBM n to denote the datasets for n universities of LUBM and UOBM, respectively. We extracted the relevant DL fragment from those ontologies, i.e. we eliminated axioms not in DL-Lite$_{core}^{\mathcal{HU}}$.

---

[2] https://github.com/kieen/OrarHSHOIF.
[3] http://sws.ifi.uio.no/project/npd-v2.
[4] https://www.cs.ox.ac.uk/isg/tools/PAGOdA.
[5] https://sites.google.com/site/ontopiswc13/home/imdb-mo.

Table 2. Test ontologies with the number of TBox axioms (# ax.), atomic concepts (# con.), roles (# rol.), individuals (# ind.), concept and role assertions (# ast.), inferred assertions (# inferred ast.) by our system

Ontology	# ax.	# con.	# rol.	# indiv.	# assert.	# inferred assert.
NPD	354	208	90	785 656	1 392 196	1 517 844
DBPedia$^+$	1 748	442	806	3 822 351	27 094 909	30 239 281
IMDb	131	88	39	6 505 584	27 757 894	33 769 170
LUBM 10	80	43	25	207 426	850 433	1 086 472
LUBM 50	80	43	25	1 082 818	4 445 949	5 676 226
LUBM 100	80	43	25	2 179 766	8 954 615	11 434 996
LUBM 500	80	43	25	10 847 183	44 573 624	56 914 960
UOBM 10	110	69	35	242 491	1 926 897	2 324 962
UOBM 50	110	69	35	1 227 123	9 751 681	11 768 772
UOBM 100	110	69	35	2 461 347	19 571 755	23 617 264
UOBM 500	110	69	35	12 375 804	98 374 692	118 717 591

Table 2 presents detailed information about the test ontologies with the number of TBox axioms, atomic concepts, roles, individuals, and (inferred) assertions. NPD, IMDb, and LUBM are in DL-Lite$_{core}^{\mathcal{H}}$ while DBPedia$^+$ and UOBM are in DL-Lite$_{core}^{\mathcal{H}\sqcup}$.

We used Orar to check consistency and compute the concept materialization of the test ontologies and compared the reasoning time of Orar and of the other well-known reasoners HermiT 1.3.8, JFact 5.0.0, Pellet 2.3.6, and Konclude 0.6.2. All tests were run on an Intel Xeon E5-2660V3 2.60 GHz machine with 250 GB heap size for the Java VM and with a timeout of five hours. Table 3 presents information about the abstractions and the size of the abstract ABoxes in comparison with the size of the original ABoxes. In NPD, IMDb, and LUBM, many individuals have the same types. For those ontologies, the size of the original ABoxes are reduced by up to four orders of magnitude. Particularly, for LUBM the abstract ABoxes are of nearly constant size regardless of the sizes of the original ABoxes. This can be explained by the simple patterns used to generate data in LUBM. The individuals in DBPedia$^+$ and UOBM are more diverse. For DBPedia$^+$, the number of types is relatively large due to the large number of concepts and roles; the size of the abstract ABox is approximately 10 % of the original one. For UOBM, the sizes of the abstract ABoxes are approximately 6 % and 1 % of the sizes of the original ones for UOBM 10 and UOBM 500, respectively. Table 4 shows the reasoning time of Orar (with Konclude as the internal reasoner for the abstraction) in comparison with the reasoning time of the other reasoners. In general, the reasoning time correlates with the size reduction of

Table 3. Number of types, abstract individuals, assertions, and size of the abstract ABox in comparison with the original ABox

Ontology	Abstraction			% of Original ABox	
	# types	# indiv.	# assert.	% indiv.	% assert.
NPD	1 005	15 580	18 244	1.983	1.310
DBPedia$^+$	226 530	1 775 630	2 770 261	46.454	10.224
IMDb	438	1 224	1 692	0.019	0.006
LUBM 10	29	154	158	0.074	0.019
LUBM 50	27	148	152	0.014	0.003
LUBM 100	27	148	152	0.007	0.002
LUBM 500	27	148	152	0.001	0.001
UOBM 10	11 391	97 944	124 661	40.391	6.470
UOBM 50	25 541	225 420	289 762	18.370	2.971
UOBM 100	34 872	310 513	400 938	12.616	2.049
UOBM 500	64 903	593 539	769 843	4.796	0.783

the ontologies. For concept materialization, Orar outperforms the other reasoners on all ontologies. For consistency checking, Konclude is faster than Orar for IMDb and LUBM. The reason is that reasoning on those ontologies is even faster than other operations required in Orar like computing types and generating the abstract ABoxes. For the other ontologies, Orar outperforms all reasoners. Note that the purpose of our evaluation was not to show the superiority of Orar, but to demonstrate that our approach can improve the performance of any existing reasoner when handling large data. Although we used Konclude inside Orar, it can be replaced by any reasoner.

6 Extensions and Variations

In this section, we discuss the extension of the presented approach to $\text{DL-Lite}_{core}^{\mathcal{HOU}}$ and present a variation of the abstract ABox, which can also be used in our approach.

6.1 Reasoning with Nominals

Since Lemma 1 even holds for the very expressive language $\mathcal{SROIQ}$, Algorithms 1 and 2 are sound for $\text{DL-Lite}_{core}^{\mathcal{HOU}}$. Before we show that they are also complete for $\text{DL-Lite}_{core}^{\mathcal{HOU}}$, we first illustrate the advantage of using the abstraction-based approach for classification of $\text{DL-Lite}_{core}^{\mathcal{HOU}}$ ontologies. This reasoning task has not been covered so far as for $\text{DL-Lite}_{core}^{\mathcal{HU}}$ classification requires reasoning only over the TBox (after checking consistency of the ontology).

Table 4. Reasoning time (without loading time) in seconds, where "−" stands for timeout

Ontology	Concept materialization					Consistency checking				
	Orar	Konclude	Pellet	HermiT	JFact	Orar	Konclude	Pellet	HermiT	JFact
NPD	5	11	39	579	−	3	8	27	284	−
DBPedia$^+$	163	176	631	2 029	−	48	148	417	292	−
IMDb	34	220	775	983	−	30	7	684	568	−
LUBM 10	2	4	9	9	3 651	2	1	8	6	2 606
LUBM 50	10	28	53	61	−	10	2	52	34	−
LUBM 100	18	67	149	133	−	16	3	135	94	−
LUBM 500	90	359	1 601	979	−	80	10	1 476	642	−
UOBM 10	8	18	23	−	−	3	16	16	47	3 073
UOBM 50	25	106	148	−	−	13	90	115	624	−
UOBM 100	42	227	353	−	−	23	187	274	1 421	−
UOBM 500	160	1 636	3 846	−	−	117	1 225	3 287	−	−

Classification of an ontology containing nominals requires reasoning over both TBox and ABox. This even holds for rather simple languages such as DL-Lite$_{core}^{\mathcal{HOU}}$ and even OWL 2 RL where nominals can only occur in a restricted form ($\exists R.o$) [14]. The following example demonstrates that class subsumption between two concepts might depend on the existence of some assertions.

Example 8. Consider a TBox $\mathcal{T} = \{A \sqsubseteq o, \exists R^- \sqsubseteq o, \exists R^- \sqsubseteq B, C \sqsubseteq \exists R\}$. We observe that $A \sqsubseteq B$ holds depending on the existence of instances of the role R, which can be enforced if C has some instances. Indeed, consider $\mathcal{A} = \{C(a)\}$, we have $\mathcal{A} \cup \mathcal{T} \models A \sqsubseteq B$. In any interpretation $\mathcal{I}$ with $A^{\mathcal{I}} = \emptyset$ the subsumption trivially holds. If, however, there is some element $d \in A^{\mathcal{I}}$, we show that d must also be in $B^{\mathcal{I}}$. By $A \sqsubseteq o$, we get $d \in o^{\mathcal{I}}$. Since $C(a) \in \mathcal{A}$, $C \sqsubseteq \exists R \in \mathcal{T}$, and $a^{\mathcal{I}} \in C^{\mathcal{I}}$, there is some d' such that $\langle a^{\mathcal{I}}, d' \rangle \in R^{\mathcal{I}}$. Since $\exists R^- \sqsubseteq o, \exists R^- \sqsubseteq B \in \mathcal{T}$, $d' \in o^{\mathcal{I}} \cap B^{\mathcal{I}}$ and, since o is a nominal concept, we have $d = d' \in A^{\mathcal{I}} \cap B^{\mathcal{I}}$ and the subsumption also holds. It is easy to see, however, that $\emptyset \cup \mathcal{T} \not\models A \sqsubseteq B$.

Since the abstractions are often smaller than the original ABoxes, classification over the abstraction will be more efficient than classification over the original ontology.

Lemma 6. *Let $\mathcal{A} \cup \mathcal{T}$ be a DL-Lite$_{core}^{\mathcal{HOU}}$ ontology and $\mathcal{B}$ the abstraction of $\mathcal{A}$. Then, for every atomic concepts $A, B \in \mathsf{con}(\mathcal{A} \cup \mathcal{T})$, $\mathcal{A} \cup \mathcal{T} \models A \sqsubseteq B$ iff $\mathcal{B} \cup \mathcal{T} \models A \sqsubseteq B$.*

By Lemmas 1 and 2, the "only-if" direction of Lemma 6 holds. We now briefly show the "if" direction of Lemma 6; and also show that Algorithms 1 and 2 are complete for DL-Lite$_{core}^{\mathcal{HOU}}$. We rely on the following extension of Lemma 4.

Lemma 7. *Let $\mathcal{O} = \mathcal{A} \cup \mathcal{T}$ be a DL-Lite$_{core}^{\mathcal{HOU}}$ ontology, $\mathcal{B}$ the abstraction of $\mathcal{A}$, and $\mathcal{B}^+ = \mathcal{B} \cup \{R(v_{\tau(a)}, v_{\tau(b)}) \mid R(a, b) \in \mathcal{A}\}$. Then, we have:*

(1) $\mathcal{B} \cup \mathcal{T}$ is consistent implies $\mathcal{B}^+ \cup \mathcal{T}$ is consistent;

(2) for every atomic concept A and individual v, $\mathcal{B}^+ \cup \mathcal{T} \models A(v)$ implies $\mathcal{B} \cup \mathcal{T} \models A(v)$; and

(3) for every atomic concepts A and B, $\mathcal{B}^+ \cup \mathcal{T} \models A \sqsubseteq B$ implies $\mathcal{B} \cup \mathcal{T} \models A \sqsubseteq B$.

Proof (Sketch). Intuitively, we follow similar steps as in the proof of Lemma 4. The only difference is to extend the interpretations to cover nominals. Reconsider the interpretations $\mathcal{I}$ and $\mathcal{J}$ in the proof of Lemma 4. We define $\mathcal{J}$ as before and let the interpretations of nominals in $\mathcal{J}$ and in $\mathcal{I}$ be identical. Then, all claims in the existing proof remain sound. Furthermore, since the interpretations of atomic concepts in $\mathcal{I}$ and in $\mathcal{J}$ are identical, we have $\mathcal{B}^+ \cup \mathcal{T} \models A \sqsubseteq B$ implies $A^{\mathcal{J}} \subseteq B^{\mathcal{J}}$, which implies $A^{\mathcal{I}} \subseteq B^{\mathcal{I}}$, i.e. $\mathcal{I} \models A \sqsubseteq B$. Since $\mathcal{I}$ is arbitrary, we have $\mathcal{B} \cup \mathcal{T} \models A \sqsubseteq B$. □

As shown in the proof of Lemma 5, there exists a homomorphism h from $\mathcal{A}$ to $\mathcal{B}^+$ that maps a to $v_{\tau(a)}$ for each $a \in \mathsf{ind}(\mathcal{A})$. By Lemmas 7 and 1, it follows that the "if" direction of the Lemma 6 holds and that Lemma 5 holds also for DL-Lite$_{core}^{\mathcal{HOU}}$.

6.2 Alternative Abstraction

The key idea of the abstraction-based approach is to build a suitable, ideally small abstract ABox, which can be used to obtain sound and complete entailments of the input ontology. The abstract ABoxes from Definition 3 induce homomorphisms to the original ABox but not necessarily vice versa. This directly guarantees soundness but not completeness. Completeness of the approach is guaranteed by Lemmas 4 and 5, which show that the abstract ABox $\mathcal{B}$ entails exactly the same assertions as its extension $\mathcal{B}^+$, to which there is a homomorphism from the original ABox. This suggests an alternative definition of abstractions similar to the extension $\mathcal{B}^+$ of $\mathcal{B}$ in Lemma 4.

Definition 5. *The* abstraction *of an ABox $\mathcal{A}$ is an ABox $\mathcal{C} = \{A(v_{\tau(a)}) \mid A(a) \in \mathcal{A}\} \cup \{R(v_{\tau(a)}, v_{\tau(b)}) \mid R(a, b) \in \mathcal{A}\}$, where $\tau(a)$ and $\tau(b)$ are the types of a and b, respectively, and $v_{\tau(a)}$ and $v_{\tau(b)}$ are a fresh, distinguished abstract individuals.*

Example 9. Consider the ABox $\mathcal{A} = \{A(a), A(b), R(a, b)\}$ as in Example 1. The abstraction of $\mathcal{A}$ by Definition 5 is the ABox $\mathcal{C} = \{A(v_{\tau(a)}), A(v_{\tau(b)}), R(v_{\tau(a)}, v_{\tau(b)})\}$.

In Example 9 there are homomorphisms both from $\mathcal{A}$ to $\mathcal{C}$ and from $\mathcal{C}$ to $\mathcal{A}$; $\mathcal{C}$ is just a copy under renaming of $\mathcal{A}$. Therefore, it is easy to see that using $\mathcal{C}$ as an abstract ABox, we obtain both sound and complete entailments for $\mathcal{A}$ w.r.t. any TBox. In general, there is always a homomorphism from $\mathcal{A}$ to $\mathcal{C}$, e.g. the mapping h defined as $h(a) = v_{\tau(a)}, a \in \mathsf{ind}(\mathcal{A})$, but not necessarily a homomorphism from $\mathcal{C}$ to $\mathcal{A}$. This immediately guarantees completeness of the approach but not soundness. However, based on our previously shown results, we can show that all results we obtained using $\mathcal{C}$ are sound. Let $\mathcal{A}$ be an ABox, $\mathcal{C}$ the abstraction of $\mathcal{A}$ by Definition 5, $\mathcal{B}$ the abstraction of $\mathcal{A}$ by Definition 3, and $\mathcal{B}^+$ the extension of $\mathcal{B}$ defined in Lemma 4. Since $\mathcal{C} \subseteq \mathcal{B}^+$, by monotonicity, for every concept A and individual $v_{\tau(a)} \in \mathsf{ind}(\mathcal{C})$, we obtain that $\mathcal{C} \cup \mathcal{T} \models A(v_{\tau(a)})$ implies $\mathcal{B}^+ \cup \mathcal{T} \models A(v_{\tau(a)})$, which implies $\mathcal{B} \cup \mathcal{T} \models A(v_{\tau(a)})$ by Lemma 4. Furthermore, by Lemma 3 we have $\mathcal{B} \cup \mathcal{T} \models A(v_{\tau(a)})$ implies $\mathcal{A} \cup \mathcal{T} \models A(a)$. Therefore, we have $\mathcal{C} \cup \mathcal{T} \models A(v_{\tau(a)})$ implies $\mathcal{A} \cup \mathcal{T} \models A(a)$.

In Definition 5, the abstract ABox $\mathcal{C}$ uses just one individual v_τ for each type τ, and, therefore, it requires less individuals than the abstract ABox $\mathcal{B}$ in Definition 3. However, for each type τ and each role R occurring in τ there is exactly one role assertion, e.g. $R(v_\tau, w_\tau^R)$, in $\mathcal{B}$, whereas v_τ could have many R-successors/predecessors in $\mathcal{C}$. In our experiment with both types of abstract ABoxes, the abstract ABoxes constructed according to Definition 5 are often larger than the ones using Definition 3.

7 Related Work

Several ontology reasoning techniques have been proposed to handle large data. The RDFox [15] and WebPIE [18] systems utilize parallel computing to perform a rule-based materialization for OWL 2 RL. The PAGOdA system [20] approximates the TBox and then performs OWL 2 RL rules to compute lower-bound and upper-bound entailments, which help to determine entailments for individuals quickly. Wandelt and Möller propose a technique for instance retrieval based on modularization [19]. A closely related work to our approaches is the SHER approach [4]. It merges individuals to obtain a compressed, *summary* ABox, which is then used for (refutation-based) consistency checking or query answering. Since merging is only based on concept assertions, the resulting summary ABox is an over-approximation of the original ABox. Therefore, if the summary ABox is consistent, then so is the original ABox, but not vice versa. In case the summary ABox is inconsistent, explanation techniques [11] are used to repair the summary. In contrast to the summary approach, the abstract ABox created in the presented approach immediately allows for both sound and complete results. Note that for DL-Lite ontologies, the previous abstraction refinement approach [7] performs reasoning over the abstract ABox twice as it continues doing refinement after transferring the entailments from the first abstraction to the original ABox.

8 Conclusions

We have presented a scalable abstraction-based approach for reasoning in DL-Lite$_{core}^{\mathcal{HU}}$ and its extensions for DL-Lite$_{core}^{\mathcal{HOU}}$. For DL-Lite$_{core}^{\mathcal{HU}}$, we focus on concept materialization and consistency checking of the ontology as computing the role materialization can be done simply by expanding the existing role assertions according to the role hierarchy; and computing the class hierarchy requires only the TBox (after checking consistency of the ontology). For DL-Lite$_{core}^{\mathcal{HOU}}$, we show that the presented approach for concept materialization and consistency checking remains sound and complete. Furthermore, it can be easily extended to ontology classification, a non-trivial task in DL-Lite$_{core}^{\mathcal{HOU}}$. Computing the role materialization of DL-Lite$_{core}^{\mathcal{HOU}}$ ontologies is not as simple as in DL-Lite$_{core}^{\mathcal{HU}}$ as role assertions can be derived not only from the role hierarchy but also from axioms of nominals. It is possible to use the abstraction to obtain also the role assertions as presented in our recent work for Horn $\mathcal{SHOIF}$ [8].

The languages we consider in this paper do not make the *Unique Name Assumption* (UNA), which is often adopted in DL-Lite. But the presented approach also works for DL-Lite$_{core}^{\mathcal{HU}}$ with UNA; for DL-Lite$_{core}^{\mathcal{HOU}}$, it does not make sense to adopt UNA. As noted in the work about different dialects of the DL-Lite family [1], we can construct a model for a DL-Lite$_{core}^{\mathcal{HU}}$ ontology with UNA from a model of that ontology without UNA by "cloning" the domain elements so that different individuals are interpreted differently. Entailments are preserved in the resulting model, therefore, the results for DL-Lite$_{core}^{\mathcal{HU}}$ without UNA remain valid in DL-Lite$_{core}^{\mathcal{HU}}$ with UNA.

References

1. Artale, A., Calvanese, D., Kontchakov, R., Zakharyaschev, M.: The DL-Lite family and relations. J. Artif. Intell. Res. **36**, 1–69 (2009)
2. Bienvenu, M., Kikot, S., Kontchakov, R., Podolskii, V.V., Zakharyaschev, M.: Theoretically optimal datalog rewritings for OWL 2 QL ontology-mediated queries. In: Proceedings of the 29th International Workshop on Description Logics (DL 2016) (2016)
3. Calvanese, D., De Giacomo, G., Lembo, D., Lenzerini, M., Rosati, R.: Tractable reasoning and efficient query answering in description logics: the DL-Lite family. J. Autom. Reason. **39**(3), 385–429 (2007)
4. Dolby, J., Fokoue, A., Kalyanpur, A., Schonberg, E., Srinivas, K.: Scalable highly expressive reasoner (SHER). J. Web Semant. **7**(4), 357–361 (2009)
5. Eiter, T., Ortiz, M., Simkus, M., Tran, T.K., Xiao, G.: Query rewriting for Horn-$\mathcal{SHIQ}$ plus rules. In: Proceedings of the 26th National Conference on Artificial Intelligence (AAAI 2012) (2012)
6. Feier, C., Carral, D., Stefanoni, G., Cuenca Grau, B., Horrocks, I.: The combined approach to query answering beyond the OWL 2 profiles. In: Proceedings of the 24th International Joint Conference on Artificial Intelligence (IJCAI 2015), pp. 2971–2977 (2015)
7. Glimm, B., Kazakov, Y., Liebig, T., Tran, T.-K., Vialard, V.: Abstraction refinement for ontology materialization. In: Mika, P., et al. (eds.) ISWC 2014, Part II. LNCS, vol. 8797, pp. 180–195. Springer, Heidelberg (2014)
8. Glimm, B., Kazakov, Y., Tran, T.: Ontology materialization by abstraction refinement in Horn $\mathcal{SHOIF}$. In: Proceedings of the 29th International Workshop on Description Logics (DL 2016) (2016)
9. Gottlob, G., Orsi, G., Pieris, A.: Query rewriting and optimization for ontological databases. ACM Trans. Database Syst. **39**(3), 25:1–25:46 (2014)
10. Horrocks, I., Kutz, O., Sattler, U.: The even more irresistible $\mathcal{SROIQ}$. In: Proceedings of 10th International Conference on Principles of Knowledge Representation and Reasoning (KR 2006), pp. 57–67. AAAI Press (2006)

11. Kalyanpur, A., Parsia, B., Horridge, M., Sirin, E.: Finding all justifications of OWL DL entailments. In: Aberer, K., et al. (eds.) ASWC 2007 and ISWC 2007. LNCS, vol. 4825, pp. 267–280. Springer, Heidelberg (2007)

12. Kikot, S., Kontchakov, R., Podolskii, V., Zakharyaschev, M.: Exponential lower bounds and separation for query rewriting. In: Czumaj, A., Mehlhorn, K., Pitts, A., Wattenhofer, R. (eds.) ICALP 2012, Part II. LNCS, vol. 7392, pp. 263–274. Springer, Heidelberg (2012)

13. Kontchakov, R., Lutz, C., Toman, D., Wolter, F., Zakharyaschev, M.: The combined approach to query answering in DL-Lite. In: Proceedings of the 12th International Conference on the Principles of Knowledge Representation and Reasoning (KR 2010). AAAI Press (2010)

14. Krötzsch, M.: The not-so-easy task of computing class subsumptions in OWL RL. In: Cudré-Mauroux, P., et al. (eds.) ISWC 2012, Part I. LNCS, vol. 7649, pp. 279–294. Springer, Heidelberg (2012)

15. Motik, B., Nenov, Y., Piro, R., Horrocks, I., Olteanu, D.: Parallel materialisation of datalog programs in centralised, main-memory RDF systems. In: Proceedings of the 28th National Conference on Artificial Intelligence (AAAI 2014), pp. 129–137 (2014)

16. Pérez-Urbina, H., Motik, B., Horrocks, I.: Tractable query answering and rewriting under description logic constraints. J. Appl. Log. 8(2), 186–209 (2010)

17. Trivela, D., Stoilos, G., Chortaras, A., Stamou, G.: Optimising resolution-based rewriting algorithms for OWL ontologies. J. Web Seman. 33, 30–49 (2015)

18. Urbani, J., Kotoulas, S., Maassen, J., van Harmelen, F., Bal, H.E.: WebPIE: a web-scale parallel inference engine using MapReduce. J. Web Seman. 10, 59–75 (2012)

19. Wandelt, S., Möller, R.: Towards ABox modularization of semi-expressive description logics. J. Appl. Ontol. 7(2), 133–167 (2012)

20. Zhou, Y., Cuenca Grau, B., Nenov, Y., Kaminski, M., Horrocks, I.: PAGOdA: pay-as-you-go ontology query answering using a datalog reasoner. J. Artif. Intell. Res. 54, 309–367 (2015)

The Impact of Active Domain Predicates on Guarded Existential Rules

Georg Gottlob[1], Andreas Pieris[2(✉)], and Mantas Šimkus[2]

[1] Department of Computer Science, University of Oxford, Oxford, UK
`georg.gottlob@cs.ox.ac.uk`
[2] Institute of Information Systems, TU Wien, Vienna, Austria
`{pieris,simkus}@dbai.tuwien.ac.at`

Abstract. We claim it is realistic to assume that a database management system provides access to the active domain via built-in relations. Therefore, product databases, i.e., databases that include designated predicates that hold the active domain, form a natural notion that deserves our attention. An important issue then is to look at the consequences of product databases for the expressiveness and complexity of central existential rule languages. We focus on guarded existential rules, and we investigate the impact of product databases on their expressive power and complexity. We show that the queries expressed via (frontier-)guarded rules gain in expressiveness, and in fact, they have the same expressive power as Datalog. On the other hand, there is no impact on the expressiveness of the queries specified via weakly-(frontier-)guarded rules since they are powerful enough to explicitly compute the predicates needed to access the active domain. We also observe that there is no impact on the complexity of the languages in question.

1 Introduction

Rule-based languages lie at the core of databases and knowledge representation. In database applications they are usually employed as query languages that go beyond standard SQL, while in knowledge representation are used for declarative problem solving, and, more recently, to model and reason about ontological knowledge. Therefore, rule-based languages can be used in at least two different ways: as query languages and as ontology languages. In the database setting, a rule-based query is expressed as a pair of the form (Σ, Ans), where Σ is a set of rules encoding the actual query, and Ans is the so-called goal predicate that collects the answer to the query. On the other hand, in the ontological setting, a database D and a set of rules Σ are used to specify implicit domain knowledge – the pair (D, Σ) is called *knowledge base* – while user queries, typically expressed as standard conjunctive queries, are evaluated over a knowledge base. Alternatively, the set of rules can be conceived as part of the specification of a query that is executed over a plain database. Such queries are known as *ontology-mediated queries* [6], and are in fact pairs of the form (Σ, q), where Σ is a set of rules expressed in a certain ontology language, and q is a conjunctive query.

© Springer International Publishing Switzerland 2016
M. Ortiz and S. Schlobach (Eds.): RR 2016, LNCS 9898, pp. 94–110, 2016.
DOI: 10.1007/978-3-319-45276-0_8

From the above discussion, it is apparent that rule-based languages form the building block of several database and ontology-mediated query languages that can be found in the literature.

An important issue for a query language (either a database or an ontology-mediated query language) is to understand its expressive power, and in particular, its expressiveness relative to other query languages. *Relative expressiveness* considers if, given two query languages L_1 and L_2, every query formulated in L_1 can be expressed by means of L_2 (and vice versa). This helps the user to choose, among a plethora of different query languages, the one that is more appropriate for the application in question. The goal of this work is to perform such an expressivity analysis for central query languages based on existential rules.

Existential rules (a.k.a. *tuple-generating dependencies* or *Datalog$^\pm$ rules*) are first-order sentences of the form $\forall \bar{x} \forall \bar{y} (\phi(\bar{x}, \bar{y}) \rightarrow \exists \bar{z}\, \psi(\bar{x}, \bar{z}))$, where ϕ and ψ are conjunctions of atoms. Intuitively speaking, such a rule states that the existence of certain tuples in a database implies the existence of some other tuples in the same database. It is widely known that the query languages based on arbitrary existential rules, without posing any syntactic restriction, are undecidable; see, e.g., [5,7]. This has led to a flurry of activity for identifying expressive fragments of existential rules that give rise to decidable query languages. One of the key paradigms that has been thoroughly studied is guardedness [2,7]. In a nutshell, the existential rule given above is guarded (resp., frontier-guarded) if ϕ has an atom that contains (or "guards") all the variables in $\bar{x} \cup \bar{y}$ (resp., $\bar{x}$). More refined languages based on weak-(frontier-)guardedness also exist.

The relative expressiveness of the languages based on (weakly-)(frontier-) guarded existential rules has been recently investigated in [9]. However, the thorough analysis performed in [9] has made no assumption on the input databases over which the queries will be evaluated, and it is known that such assumptions may have an impact on the expressiveness of a query language. Recall the classical result that semipositive Datalog over ordered databases is powerful enough to express all queries that are computable in polynomial time, which is not true without assuming ordered databases [1].

We claim it is natural to focus on *product databases*, that is, databases that include designated predicates that hold the active domain. In other words, those predicates give access to the cartesian product of the active domain (hence the name product databases). Since it is realistic to assume that a database management system provides access to the active domain via built-in relations (e.g., lookup or reference tables), we believe that product databases form a central notion that deserves our attention. In view of this fact, it is important to understand how the relative expressiveness of the guarded-based query languages in question is affected when we concentrate on product databases. This is the goal of the present work. The outcome of our analysis can be summarized as follows:

- The query languages based on (frontier-)guarded existential rules gain in expressiveness, and, in fact, they have the same expressive power as Datalog.

– There is no impact on the expressive power of the query languages that are based on weakly-(frontier-)guarded existential rules, since they are powerful enough to explicitly compute the relations needed to access the active domain.
– Finally, we show that there is no impact on the computational complexity of the guarded-based query languages in question.

Although the employed techniques for establishing the above results are rather standard, which build on existing ones that can be found in the literature, the obtained results are conceptually interesting (e.g., assuming product databases, (frontier-)guardedness gives rise to query languages that are equally expressive to Datalog). We believe that our analysis sheds light on the expressivity of the guarded-based query languages in question, and complements the recent investigation preformed in [9]. Let us clarify that in the above summarization of our results, the term query language refers to both database and ontology-mediated query languages. Since the former is a special case of the latter (indeed, the query $(\Sigma, \mathtt{Ans})$ is actually the ontology-mediated query $(\Sigma, \mathtt{Ans}(x_1, \ldots, x_n))$, where n is the arity of $\mathtt{Ans}$), in the sequel we focus on ontology-mediated queries.

2   Motivating Example

The goal of this section is to illustrate, via a meaningful example, that product databases have an impact on the expressiveness of frontier-guarded ontology-mediated queries, which in turn allows us to write complex queries in a more flexible way. Suppose we are developing a system for managing a response to a natural disaster. The ultimate goal of the system is to collect information about volunteers and their qualifications, and then use this information to coordinate various relief activities.

The Database. Suppose that the database of such a system contains a binary relation Team that stores an assignment of volunteers to teams. For example, the atom

$$\mathtt{Team}(\text{``Alpha''}, \text{``Ann''})$$

means that Ann belongs to the team called Alpha. The database also includes a binary relation called ExperienceIn, which relates persons to tasks in which they have experience. For instance, the atom

$$\mathtt{ExperienceIn}(\text{``John''}, \text{``perform CPR''})$$

states that John has experience in performing CPR. We also have a binary relation hasTraining with the obvious meaning; for example,

$$\mathtt{hasTraining}(\text{``John''}, \text{``race driver''})$$

means that John has been trained to drive a race car. In addition, the database contains a unary relation ProDriverQualification that stores qualifications that involve driving at professional level; e.g.,

$$\mathtt{ProDriverQualification}(\text{``bus license''})$$

states that bus license is a qualification to drive at professional level. We further assume that some tasks that can be performed by volunteers are grouped into more complex procedures. For instance, the response to a water leak could consist of performing four tasks in the following order: load equipment, drive truck, perform repairs and clean up. This is stated in the database of the system using the atoms:

$$\texttt{ProcedureTaskFirst}(\text{``water leak''}, \text{``load equipment''})$$
$$\texttt{ProcedureTaskOrder}(\text{``water leak''}, \text{``load equipment''}, \text{``drive truck''})$$
$$\texttt{ProcedureTaskOrder}(\text{``water leak''}, \text{``drive truck''}, \text{``perform repairs''})$$
$$\texttt{ProcedureTaskOrder}(\text{``water leak''}, \text{``perform repairs''}, \text{``clean up''})$$
$$\texttt{ProcedureTaskLast}(\text{``water leak''}, \text{``clean up''}).$$

Intuitively, $\texttt{ProcedureTaskFirst}(p, t)$ and $\texttt{ProcedureTaskLast}(p, t')$ state that t/t' are the first/last task in the procedure p. The atom $\texttt{Procedure} \texttt{TaskOrder}(p, t, t')$ says that in the procedure p the task t' follows the task t.

The Ontology. We know that some intensional knowledge, not explicitly stored in the database described above, also holds. More precisely, we know that if a person p has experience in some task t, then p is qualified to perform t. This can be expressed as

$$\sigma_1 = \texttt{ExperienceIn}(Pn, Tk) \; \rightarrow \; \texttt{QualifiedFor}(Pn, Tk).$$

Moreover, we know that if a person p has been trained to be a professional driver, then p is qualified to drive an ambulance. This can be expressed as

$$\sigma_2 = \texttt{hasTraining}(Pn, T), \texttt{ProDriverQualification}(T) \; \rightarrow$$
$$\texttt{QualifiedFor}(Pn, \text{``drive ambulance''}).$$

In addition, if a person p is experienced in delivering heavy goods, then p must have some training that leads to a truck license. This is expressed via the rule

$$\sigma_3 = \texttt{ExperienceIn}(Pn, \text{``delivery heavy goods''}) \; \rightarrow$$
$$\exists T \, \texttt{hasTraining}(Pn, T), \texttt{TruckLicense}(T).$$

Finally, truck license leads to a professional driving license, which can be expressed as

$$\sigma_4 = \texttt{TruckLicense}(T) \; \rightarrow \; \texttt{ProDriverQualification}(T).$$

Observe that our ontology $\Sigma = \{\sigma_1, \ldots, \sigma_4\}$ consists of guarded existential rules.

The Database Query. In our disaster management scenario we are interested in checking whether a team is qualified to perform every task of a certain procedure. More precisely, we want to collect in a binary relation $\texttt{TeamQualified}$ all pairs (t, p) of a team and a procedure such that: for every task j of the procedure p, the team t has a member m that is qualified for j. Recall that an

ontology-mediated query is a pair of an ontology and a database query. Therefore, we need to express the above query as a database query q, which, together
with the ontology Σ defined above, will give rise to the ontology-mediated query
(Σ, q). Unfortunately, things are a bit more complicated than they seem. In
particular, the query q is inherently recursive, and thus is not expressible as
a conjunctive query. However, it can be easily expressed as the Datalog query
$(\Pi, \texttt{TeamQualified})$, where the program Π consists of the rules:

$$\rho_1 = \begin{array}{l} \texttt{ProcedureTaskFirst}(Pc,\, Tk), \\ \texttt{Team}(Tm,\, Pn), \\ \texttt{QualifiedFor}(Pn,\, Tk) \;\rightarrow\; \texttt{QualifiedUntil}(Tm,\, Pc,\, Tk) \end{array}$$

$$\rho_2 = \begin{array}{l} \texttt{ProcedureTaskOrder}(Pc,\, Tk',\, Tk), \\ \texttt{Team}(Tm,\, Pn), \\ \texttt{QualifiedUntil}(Tm,\, Pn,\, Tk') \;\rightarrow\; \texttt{QualifiedUntil}(Tm,\, Pc,\, Tk) \end{array}$$

$$\rho_3 = \begin{array}{l} \texttt{ProcedureTaskLast}(Pc,\, Tk), \\ \texttt{QualifiedUntil}(Tm,\, Pc,\, Tk) \;\rightarrow\; \texttt{TeamQualified}(Tm,\, Pc). \end{array}$$

The fact that our query q is expressible as a recursive Datalog query is of little
use since the ontology-mediated query (Σ, q) does not comply with the formal
definition of ontology-mediated queries where q must be a first-order query, and
thus does not fall in a decidable guarded-based ontology-mediated query language. Hence, the crucial question that comes up is whether we can construct a
query (Σ', q') that is equivalent to (Σ, q), while Σ is a set of (frontier-)guarded
existential rules and q is a conjunctive query. One may think that this can be
achieved by adding the rules of Π in the ontology Σ, i.e., $\Sigma' = \Sigma \cup \Pi$, and let
q be the atomic conjunctive query $\texttt{TeamQualified}(x, y)$. Although the obtained
query (Σ', q') is equivalent to (Σ, q), it is inherently unguarded, and it cannot
be expressed as a frontier-guarded ontology-mediated query. However, assuming
that our database is product, which gives us access to the active domain via relations of the form Dom^k, for $k > 0$, that hold all the k-tuples of constants occurring
in the active domain, we can convert the rules $\rho_1, \rho_2 \in \Sigma'$ into guarded rules,
without changing the meaning of the query (Σ', q'), by adding in their body the
atom $\mathrm{Dom}^4(Tm, Pc, Tk, Pn)$ and $\mathrm{Dom}^5(Tm, Pc, Tk, Tk', Pn)$, respectively. Hence,
the assumption that the database is product allows us to rewrite the query (Σ, q)
into an equivalent guarded ontology-mediated query.

3   Preliminaries

Instances and Queries. Let $\mathbf{C}$, $\mathbf{N}$ and $\mathbf{V}$ be pairwise disjoint countably infinite
sets of *constants*, (labeled) *nulls* and *variables* (used in queries and dependencies),
respectively. A *schema* $\mathbf{S}$ is a finite set of relation symbols (or predicates) with
associated arity. We write R/n to denote that R has arity n. A *term* is either
a constant, null or variable. An *atom* over $\mathbf{S}$ is an expression $R(\bar{t})$, where R is
a relation symbol in $\mathbf{S}$ of arity $n > 0$ and $\bar{t}$ is an n-tuple of terms. A *fact* is

an atom whose arguments consist only of constants. An *instance* over **S** is a (possibly infinite) set of atoms over **S** that contain constants and nulls, while a *database* over **S** is a finite set of facts over **S**. The *active domain* of an instance I, denoted $adom(I)$, is the set of all terms occurring in I.

A *query* over **S** is a mapping q that maps every database D over **S** to a set of *answers* $q(D) \subseteq adom(D)^n$, where $n \geq 0$ is the *arity* of q. The usual way of specifying queries is by means of (fragments of) first-order logic. Such a central fragment is the class of conjunctive queries. A *conjunctive query* (CQ) q over **S** is a conjunction of atoms of the form $\exists \bar{y}\, \phi(\bar{x}, \bar{y})$, where $\bar{x} \cup \bar{y}$ are variables of **V**, that uses only predicates from **S**. The free variables of a CQ are called *answer variables*. The evaluation of CQs over instances is defined in terms of homomorphisms. A *homomorphism* from a set of atoms A to a set of atoms A' is a partial function $h : \mathbf{C} \cup \mathbf{N} \cup \mathbf{V} \rightarrow \mathbf{C} \cup \mathbf{N} \cup \mathbf{V}$ such that: (i) $t \in \mathbf{C}$ implies $h(t) = t$, i.e., is the identity on **C**, and (ii) $R(t_1, \ldots, t_n) \in A$ implies $h(R(t_1, \ldots, t_n)) = R(h(t_1), \ldots, h(t_n)) \in A'$. The *evaluation* of q over an **S**-instance I, denoted $q(I)$, is the set of all tuples $h(\bar{x})$ of constants such that h is a homomorphism from q to I. Each schema **S** and CQ $q = \exists \bar{y}\, \phi(x_1, \ldots, x_n, \bar{y})$ give rise to the n-ary query $q_{\phi, \mathbf{S}}$ defined by setting, for every database D over **S**, $q_{\phi, \mathbf{S}}(D) = \{\bar{c} \in adom(D)^n \mid \bar{c} \in q(D)\}$. Let CQ be the class of all queries definable by some CQ.

Tgds for Specifying Ontologies. An ontology language is a fragment of first-order logic. We focus on ontology languages that are based on tuple-generating dependencies. A *tuple-generating dependency* (tgd) is a first-order sentence of the form

$$\forall \bar{x} \forall \bar{y} \big(\phi(\bar{x}, \bar{y}) \; \rightarrow \; \exists \bar{z}\, \psi(\bar{x}, \bar{z}) \big),$$

where both ϕ and ψ are conjunctions of atoms without nulls and constants. For simplicity, we write this tgd as $\phi(\bar{x}, \bar{y}) \rightarrow \exists \bar{z}\, \psi(\bar{x}, \bar{z})$, and use comma instead of "$\wedge$" for conjoining atoms. We call ϕ and ψ the *body* and *head* of the tgd, respectively, and write $sch(\Sigma)$ for the set of predicates occurring in Σ. An instance I *satisfies* the above tgd if: For every homomorphism h from $\phi(\bar{x}, \bar{y})$ to I, there is a homomorphism h' that extends h, i.e., $h' \supseteq h$, from $\psi(\bar{x}, \bar{z})$ to I. I satisfies a set Σ of tgds, denoted $I \models \Sigma$, if I satisfies every tgd in Σ. Let TGD be the class of all (finite) sets of tgds.

Ontology-Mediated Queries. An *ontology-mediated query* is a triple $(\mathbf{S}, \Sigma, q)$, where **S** is a schema, called *data schema*, $\Sigma \in$ TGD, $q \in$ CQ, and q is over $\mathbf{S} \cup sch(\Sigma)$.[1] Notice that the data schema **S** is included in the specification of an ontology-mediated query in order to make clear that the query is over **S**, i.e., it ranges over **S**-databases. The semantics of such a query is defined in terms of certain answers. Let $(\mathbf{S}, \Sigma, q)$ be an ontology-mediated query, where n is the arity of q. The *answer* to q with respect to a database D over **S** and Σ is $cert_{q, \Sigma}(D) = \bigcap_{I \supseteq D, I \models \Sigma} \{\bar{c} \in adom(D)^n \mid \bar{c} \in q(I)\}$.

---

[1] In fact, ontology-mediated queries can be defined for arbitrary ontology and query languages.

At this point, it is important to recall that $cert_{q,\Sigma}(D)$ coincides with the evaluation of q over the canonical instance of D and Σ that can be constructed by applying the chase procedure [7,8,10,11]. Roughly speaking, the chase adds new atoms to D as dictated by Σ until the final result satisfies Σ, while the existentially quantified variables are satisfied by inventing fresh null values. The formal definition of the chase procedure follows. Let I be an instance and $\sigma = \phi(\bar{x}, \bar{y}) \rightarrow \exists \bar{z} \psi(\bar{x}, \bar{z})$ a tgd. We say that σ is *applicable* with respect to I if there exists a homomorphism h from $body(\sigma)$ to I. In this case, *the result of applying σ over I with h* is the instance $J = I \cup h'(head(\sigma))$, where h' is an extension of h that maps each $z \in \bar{z}$ to a fresh null value not in I. For such a single chase step we write $I \xrightarrow{\sigma, h} J$. Let us assume now that I is an instance and Σ a finite set of tgds. A *chase sequence for I under Σ* is a (finite or infinite) sequence: $I_0 \xrightarrow{\sigma_0, h_0} I_1 \xrightarrow{\sigma_1, h_1} I_2 \ldots$ of chase steps such that: (1) $I_0 = I$; (2) For each $i \geq 0$, $\sigma_i \in \Sigma$; and (3) $\bigcup_{i \geq 0} I_i \models \Sigma$. Notice that in case the above chase sequence is infinite, then it must be also *fair*, that is, whenever a tgd $\sigma \in \Sigma$ is applicable with respect to I_i with homomorphism h_i, then there exists $h' \supseteq h_i$ and $k > i$ such that $h'(head(\sigma)) \subseteq I_k$. In other words, a fair chase sequence guarantees that all tgds that are applicable will eventually be applied. We call $\bigcup_{i \geq 0} I_i$ the *result* of this chase sequence, which always exists. Although the result of a chase sequence is not necessarily unique (up to isomorphism), each such result is equally useful for our purposes since is *universal*, that is, it can be homomorphically embedded into every other result. Therefore, we denote by $chase(I, \Sigma)$ *the* result of an arbitrary chase sequence for I under Σ.

Given an ontology-mediated query $(\mathbf{S}, \Sigma, q)$, it is well-known that $cert_{q,\Sigma}(D) = q(chase(D, \Sigma))$, for every $\mathbf{S}$-database D. In other words, to compute the answer to q with respect to D and Σ, we simply need to evaluate q over the instance $chase(D, \Sigma)$. Notice that this does not provide an effective algorithm for computing $cert_{q,\Sigma}(D)$ since the instance $chase(D, \Sigma)$ is, in general, infinite.

Ontology-Mediated Query Languages. Every ontology-mediated query $Q = (\mathbf{S}, \Sigma, q)$ can be interpreted as a query q_Q over $\mathbf{S}$ by setting $q_Q(D) = cert_{q,\Sigma}(D)$, for every $\mathbf{S}$-database D. Thus, we obtain a new query language, denoted $(\mathsf{TGD}, \mathsf{CQ})$, defined as the class of queries q_Q, where Q is an ontology-mediated query. However, $(\mathsf{TGD}, \mathsf{CQ})$ is undecidable since, given a database D over $\mathbf{S}$, $\Sigma \in \mathsf{TGD}$, an n-ary query $q \in \mathsf{CQ}$ over $\mathbf{S} \cup sch(\Sigma)$, and a tuple $\bar{c} \in \mathbf{C}^n$, the problem of deciding whether $\bar{c} \in cert_{q,\Sigma}(D)$ is undecidable; see, e.g., [5,7]. This has led to a flurry of activity for identifying decidable syntactic restrictions. Such a restriction defines a subclass $\mathcal{C}$ of tgds, i.e., $\mathcal{C} \subseteq \mathsf{TGD}$, which in turn gives rise to the query language $(\mathcal{C}, \mathsf{CQ})$. Such a query language is called *ontology-mediated query language*. Here we focus on ontology-mediated query languages that are based on the notion of guardedness:

(Frontier-)Guarded Tgds: A tgd is *guarded* if its body contains an atom, called *guard*, that contains all the body-variables [7]. Let G be the class of all finite sets of guarded tgds. A key extension of guarded tgds is the class of *frontier-guarded*

tgds, where the guard contains only the frontier variables, i.e., the body-variables that appear in the head [2]. Let FG be the class of all finite sets of frontier-guarded tgds.

Weak Versions: Both G and FG have a weak version: Weakly-guarded [7] and weakly-frontier-guarded [2], respectively. These are highly expressive classes of tgds obtained by relaxing the underlying condition so that only those variables that may unify with null values during the chase are taken into account. In order to formalize these classes of tgds we need some additional terminology. A *position* $R[i]$ identifies the i-th attribute of a predicate R. Given a schema $\mathbf{S}$, the set of positions of $\mathbf{S}$ is the set $\{R[i] \mid R/n \in \mathbf{S} \text{ and } i \in \{1,\ldots,n\}\}$. Given a set Σ of tgds, the set of *affected positions* of $sch(\Sigma)$, denoted $affected(\Sigma)$, is inductively defined as follows: (1) If there exists $\sigma \in \Sigma$ such that at position π an existentially quantified variable occurs, then $\pi \in affected(\Sigma)$; and (2) If there exists $\sigma \in \Sigma$ and a variable V in $body(\sigma)$ only at positions of $affected(\Sigma)$, and V appears in $head(\sigma)$ at position π, then $\pi \in affected(\Sigma)$. A tgd σ is *weakly-guarded with respect to* Σ if its body contains an atom, called *weak-guard*, that contains all the body-variables that appear only at positions of $affected(\Sigma)$. The set Σ is *weakly-guarded* if each $\sigma \in \Sigma$ is weakly-guarded with respect to Σ. The class of weakly-frontier-guarded sets of tgds is defined analogously, but considering only the body-variables that appear also in the head of a tgd. We write WG (resp., WFG) for the class of all finite weakly-guarded (resp., weakly-frontier-guarded) sets of tgds.

4   Product Databases

Recall that product databases provide access to the active domain via designated built-in predicates. Before proceeding to the next section, where we look at the impact of product databases on the expressive power of the ontology-mediated query languages in question, let us make the notion of a product database more precise.

A database D is said to be α-*product*, where α is a finite set of positive integers, if it includes a designated predicate $\mathtt{Dom}^i/i$, for each $i \in \alpha$, that holds all the i-tuples of constants in $adom(D)$, or, in other words, the restriction of D over the predicate $\mathtt{Dom}^i$ is precisely the set of facts $\{\mathtt{Dom}^i(\bar{t}) \mid \bar{t} \in adom(D)^i\}$. Given a non-product database D, we denote by D^α the α-product database $D \cup \{\mathtt{Dom}^i(\bar{t}) \mid \bar{t} \in adom(D)^i\}_{i \in \alpha}$. An *ontology-mediated query over a product database* is an ontology-mediated query $(\mathbf{S}, \Sigma, q)$ such that $\mathbf{S}$ contains the predicates $\mathtt{Dom}^{i_1},\ldots,\mathtt{Dom}^{i_k}$, for some set of positive integers $\alpha = \{i_1,\ldots,i_k\}$, while none of those predicates appears in the head of a tgd of Σ. The latter condition is posed since the predicates $\mathtt{Dom}^{i_1},\ldots,\mathtt{Dom}^{i_k}$ are conceived as built-in read-only predicates, and thus, we cannot modify their content. Such a query ranges only over $\mathbf{S}$-databases that are α-product. We write $(\mathcal{C}, \mathsf{CQ})_\times$ for the class of $(\mathcal{C}, \mathsf{CQ})$ queries over a product database.

Example 1. Consider the query $Q_{trans} = (\{E\}, \Sigma, \text{Ans}(x, y))$, where Σ is the set:

$$E(x, y) \rightarrow T(x, y)$$
$$E(x, y), T(y, z) \rightarrow T(x, z)$$
$$T(x, y) \rightarrow \text{Ans}(x, y),$$

which computes the transitive closure of the binary predicate E. It is easy to see that the above query can be equivalently rewritten as a guarded ontology-mediated query over a product database, i.e., as a $(\mathsf{G}, \mathsf{CQ})_\times$ query. More precisely, Q_{trans} can be written as $Q'_{trans} = (\{E, \text{Dom}^3\}, \Sigma', \text{Ans}(x, y))$, where Σ' is the set of tgds:

$$E(x, y) \rightarrow T(x, y)$$
$$\text{Dom}^3(x, y, z), E(x, y), T(y, z) \rightarrow T(x, z)$$
$$T(x, y) \rightarrow \text{Ans}(x, y).$$

Clearly, for every $\{E\}$-database D, $Q_{trans}(D) = Q'_{trans}(D^{\{3\}})$.   ∎

5   The Impact of Product Databases

We are now ready to investigate the impact of product databases on the relative expressiveness of the guarded-based ontology-mediated query languages in question. Let us first fix some auxiliary terminology. Two ontology-mediated queries $Q_1 = (\mathbf{S}_1, \Sigma_1, q_1)$ and $Q_2 = (\mathbf{S}_2, \Sigma_2, q_2)$ over a product database, with $\alpha_i = \{j \mid \text{Dom}^j \in \mathbf{S}_i\}$, for each $i \in \{1, 2\}$, are *comparable relative to schema* $\mathbf{S}$ if $\mathbf{S} = \mathbf{S}_1 \setminus \{\text{Dom}^i \mid i \in \alpha_1\} = \mathbf{S}_2 \setminus \{\text{Dom}^i \mid i \in \alpha_2\}$. Such comparable queries are *equivalent*, written $Q_1 \equiv Q_2$, if, for every database D over $\mathbf{S}$, $cert_{q_1, \Sigma_1}(D^{\alpha_1}) = cert_{q_2, \Sigma_2}(D^{\alpha_2})$. It is important to say that the above definitions immediately apply even if we consider queries that are not over a product database. An ontology-mediated query language $\mathcal{Q}_2$ is *at least as expressive as* the ontology-mediated query language $\mathcal{Q}_1$, written $\mathcal{Q}_1 \preceq \mathcal{Q}_2$, if, for every $Q_1 \in \mathcal{Q}_1$ there is $Q_2 \in \mathcal{Q}_2$ such that Q_1 and Q_2 are comparable (relative to some schema) and $Q_1 \equiv Q_2$. $\mathcal{Q}_2$ is *strictly more expressive than* $\mathcal{Q}_1$, written $\mathcal{Q}_1 \prec \mathcal{Q}_2$, if $\mathcal{Q}_1 \preceq \mathcal{Q}_2 \npreceq \mathcal{Q}_1$. $\mathcal{Q}_1$ and $\mathcal{Q}_2$ *have the same expressive power*, written $\mathcal{Q}_1 = \mathcal{Q}_2$, if $\mathcal{Q}_1 \preceq \mathcal{Q}_2 \preceq \mathcal{Q}_1$. In our analysis we also include Datalog. Recall that a Datalog program is simply a set of single-head tgds without existentially quantified variables, while a Datalog query over $\mathbf{S}$ of the form $(\Sigma, \text{Ans}/n)$, where Σ is a Datalog program and Ans is the answer predicate, can be seen as the ontology-mediated query $(\mathbf{S}, \Sigma, \text{Ans}(x_1, \ldots, x_n))$. We write DAT for the class of queries definable via some Datalog query. We show the following:

Theorem 1. *It holds that,*

$$(\mathsf{G}, \mathsf{CQ}) \prec (\mathsf{FG}, \mathsf{CQ}) \prec (\mathsf{G}, \mathsf{CQ})_\times = (\mathsf{FG}, \mathsf{CQ})_\times = \mathsf{DAT} \prec$$
$$(\mathsf{WG}, \mathsf{CQ}) = (\mathsf{WFG}, \mathsf{CQ}) = (\mathsf{WG}, \mathsf{CQ})_\times = (\mathsf{WFG}, \mathsf{CQ})_\times.$$

The rest of this section is devoted to establish the above result. This is done by establishing a series of technical lemmas that all together imply Theorem 1. Henceforth, we assume that, given an ontology-mediated query $(\mathbf{S}, \Sigma, q)$, none of the predicates of $\mathbf{S}$ occur in the head of a tgd of Σ. This assumption can be made without loss of generality since, for each $R \in \mathbf{S}$ that appears in the head of tgd of Σ, we can add to Σ the auxiliary copy tgd $R(x_1, \ldots, x_n) \rightarrow R^*(x_1, \ldots, x_n)$, and then replace each occurrence of R in Σ and q with R^*. We first establish that frontier-guarded ontology-mediated queries are strictly more expressive than guarded ontology-mediated queries. Although this is generally known, it is not explicitly shown in some previous work. Hence, for the sake of completeness, we would like to provide a proof sketch for this fact.

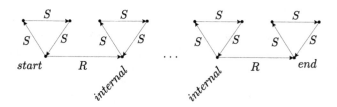

Fig. 1. The graph from the proof of Lemma 1.

Lemma 1. $(\mathsf{G}, \mathsf{CQ}) \prec (\mathsf{FG}, \mathsf{CQ})$.

Proof (sketch). We need to exhibit a query that can be expressed in $(\mathsf{FG}, \mathsf{CQ})$ but not in $(\mathsf{G}, \mathsf{CQ})$, which in turn shows that $(\mathsf{FG}, \mathsf{CQ}) \not\preceq (\mathsf{G}, \mathsf{CQ})$; the other direction holds trivially since $\mathsf{G} \subseteq \mathsf{FG}$. Such a query is the one that asks whether a labeled directed graph $G = (N, E, \lambda, \mu)$, where $\lambda : N \rightarrow \{start, internal, end\}$ and $\mu : E \rightarrow \{R, S\}$, contains a directed R-path P from a start node to an end node via internal nodes, while each node of P is part of a directed S-triangle. In other words, we ask if the graph G contains a subgraph as the one depicted in Fig. 1. The graph G is naturally encoded in an $\mathbf{S}$-database D, where $\mathbf{S} = \{\mathtt{Start}/1, \mathtt{Internal}/1, \mathtt{End}/1, R/2, S/2\}$. Our query can be expressed as the $(\mathsf{FG}, \mathsf{CQ})$ query $Q = (\mathbf{S}, \Sigma, \mathtt{Yes}())$, where Σ consists of:

$$S(x, x_1), S(x_1, x_2), S(x_2, x) \rightarrow \mathtt{Triangle}_S(x)$$
$$\mathtt{Start}(x) \rightarrow \mathtt{Mark}(x)$$
$$\mathtt{Mark}(x), \mathtt{Triangle}_S(x), R(x, y), \mathtt{Internal}(y) \rightarrow \mathtt{Mark}(y)$$
$$\mathtt{Mark}(x), \mathtt{Triangle}_S(x), R(x, y), \mathtt{End}(y), \mathtt{Triangle}_S(y) \rightarrow \mathtt{Yes}().$$

Let us intuitively explain why Q cannot be expressed as a $(\mathsf{G}, \mathsf{CQ})$ query. Assume that Q can be expressed via the $(\mathsf{G}, \mathsf{CQ})$ query $(\mathbf{S}, \Sigma', q')$. It is well-known that the query that asks whether a node belongs to an S-triangle is unguarded, and thus, it cannot be expressed via a query of the form $(\mathbf{S}, \Sigma', q_a)$, where $\Sigma' \in \mathsf{G}$ and q_a is atomic. Thus, the "triangle checks" must necessarily be performed by the CQ q'. This implies that q' can perform an unbounded number of "triangle checks", and thus, q' can express a query that is inherently recursive. But this contradicts the fact that a (finite) first-order query, let alone a conjunctive query, cannot express a recursive query. □

We proceed to show that Datalog queries are strictly more expressive than frontier-guarded ontology-mediated queries. Towards this end, we are going to exploit the fact that $(\mathsf{FG}, \mathsf{ACQ}) \preceq \mathsf{DAT}$, where ACQ is the class of queries definable by some atomic CQ [9].[2] This means that, given a query $Q = (\mathbf{S}, \Sigma, \exists \bar{y}\, \mathsf{Ans}(\bar{x}, \bar{y})) \in (\mathsf{FG}, \mathsf{ACQ})$, there exists a procedure Ξ that translates Σ into a Datalog program such that Q and the query $(\Xi(\Sigma), \mathsf{Ans}) \in \mathsf{DAT}$ over $\mathbf{S}$ are equivalent.

Lemma 2. $(\mathsf{FG}, \mathsf{CQ}) \prec \mathsf{DAT}$.

Proof (sketch). We first show that $(\mathsf{FG}, \mathsf{CQ}) \preceq \mathsf{DAT}$. Let $Q = (\mathbf{S}, \Sigma, q) \in (\mathsf{FG}, \mathsf{CQ})$, with $q = \exists \bar{y}\, \phi(x_1, \ldots, x_n, \bar{y})$. Q can be equivalently rewritten as a $(\mathsf{TGD}, \mathsf{ACQ})$ query. More precisely, Q is equivalent to the query

$$Q' = (\mathbf{S} \cup \{P, P^\star\}, \Sigma_{P^\star} \cup \Sigma \cup \{\sigma_q\}, \mathsf{Ans}(x_1, \ldots, x_n)),$$

where $P/1, P^\star/n$ are auxiliary predicates not in $\mathbf{S} \cup sch(\Sigma)$, $\Sigma_{P^\star}$ consists of the tgds:

$$R(x_1, \ldots, x_n) \to P(x_i), \text{ for each } R \in \mathbf{S} \text{ and } i \in \{1, \ldots, n\}$$
$$P(x_1), \ldots, P(x_n) \to P^\star(x_1, \ldots, x_n),$$

and σ_q is the tgd

$$P^\star(x_1, \ldots, x_n), \phi(x_1, \ldots, x_n, \bar{y}) \to \mathsf{Ans}(x_1, \ldots, x_n).$$

In particular, $\Sigma_{P^\star}$ defines the predicate $P^\star$ that holds all the n-tuples over constants of the active domain, which then can be used in σ_q that converts the CQ q into a frontier-guarded tgd. Notice that Q' is not a query over a product database, which means we do not have access to the built-in predicate Dom^n. Therefore, in order to convert q into a frontier-guarded tgd, we need to explicitly construct all the n-tuples over the active domain and store them in the auxiliary predicate $P^\star$. Although the set of tgds $\Sigma' = \Sigma_{P^\star} \cup \Sigma \cup \{\sigma_q\}$ is not frontier-guarded, it has a very special form that allows us to rewrite it into a Datalog program by applying the translation Ξ. Observe that Σ' admits a stratification, where the first stratum is the set $\Sigma_{P^\star}$, while the second stratum is the frontier-guarded set $\Sigma \cup \{\sigma_q\}$. This implies that Q' is equivalent to the Datalog query $(\Sigma_{P^\star} \cup \Xi(\Sigma \cup \{\sigma_q\}), \mathsf{Ans})$ over $\mathbf{S}$, and the claim follows.

It remains to show that $\mathsf{DAT} \not\preceq (\mathsf{FG}, \mathsf{CQ})$. To this end, it suffices to construct a Datalog query Q over a schema $\mathbf{S}$ such that, for every query $Q' \in (\mathsf{FG}, \mathsf{CQ})$ over $\mathbf{S}$, there exists an $\mathbf{S}$-database D such that, $Q(D) \neq Q'(D)$. We claim that

---

[2] A similar result can be found in [4].

such a Datalog query is Q_{trans} given in Example 1, which computes the transitive closure of the binary relation E. Towards a contradiction, assume that Q_{trans} can be expressed as $(\mathsf{FG}, \mathsf{CQ})$ query $(\mathbf{S}, \Sigma, q)$. Observe that frontier-guarded tgds are not able to put together in an atom, during the construction of the chase instance, two database constants that do not already coexist in a database atom. In particular, given a database D and a set $\Sigma \in \mathsf{FG}$, if there is no atom in D that contains the constants $c, d \in adom(D)$, then there is no atom in $chase(D, \Sigma)$ that contains c and d. Therefore, q is able to compute the transitive closure of the binary relation E. But this contradicts the fact that a (finite) conjunctive query cannot compute the transitive closure of a binary relation. $\qquad\square$

We now show that product databases have an impact on the expressiveness of the ontology-mediated query languages based on (frontier-)guarded tgds. In fact, these languages become equally expressive to Datalog when we focus on product databases.

Lemma 3. $(\mathsf{G}, \mathsf{CQ})_\times = (\mathsf{FG}, \mathsf{CQ})_\times = \mathsf{DAT}$

Proof. First observe that $(\mathsf{FG}, \mathsf{CQ})_\times = (\mathsf{FG}, \mathsf{ACQ})_\times$; recall that ACQ is the class of queries definable by some atomic CQ. More precisely, a query $(\mathbf{S}, \Sigma, q) \in (\mathsf{FG}, \mathsf{CQ})$, with $q = \exists \bar{y}\, \phi(x_1, \ldots, x_n, \bar{y})$, is equivalent to the $(\mathsf{FG}, \mathsf{ACQ})_\times$ query

$$(\mathbf{S}, \Sigma \cup \{\sigma_q\}, \mathtt{Ans}(x_1, \ldots, x_n)),$$

where σ_q is the tgd

$$\mathtt{Dom}^n(x_1, \ldots, x_n), \phi(x_1, \ldots, x_n, \bar{y}) \rightarrow \mathtt{Ans}(x_1, \ldots, x_n),$$

which implies that $(\mathsf{FG}, \mathsf{CQ})_\times \preceq (\mathsf{FG}, \mathsf{ACQ})_\times$; the other direction holds trivially. Therefore, to prove our claim, it suffices to show that

$$(\mathsf{G}, \mathsf{CQ})_\times \overset{(1)}{\preceq} (\mathsf{FG}, \mathsf{ACQ})_\times \overset{(2)}{\preceq} \mathsf{DAT} \overset{(3)}{\preceq} (\mathsf{G}, \mathsf{CQ})_\times.$$

For showing (1), we observe that the construction given above for rewriting a $(\mathsf{FG}, \mathsf{CQ})_\times$ query into a $(\mathsf{FG}, \mathsf{ACQ})_\times$ query can be used in order to rewrite a $(\mathsf{G}, \mathsf{CQ})_\times$ query into a $(\mathsf{FG}, \mathsf{ACQ})_\times$ query. For showing (2), we can apply the procedure Ξ mentioned above, which transforms a $(\mathsf{FG}, \mathsf{ACQ})$ query into an equivalent DAT query. Finally, (3) follows from the fact that a Datalog rule ρ can be converted into a guarded tgd by adding in the body of ρ the atom $\mathtt{Dom}^{|\bar{x}|}(\bar{x})$, where $\bar{x}$ are the variables in ρ. $\qquad\square$

The next lemma shows that weakly-guarded sets of tgds give rise to an ontology-mediated query language that is strictly more expressive than Datalog.

Lemma 4. $\mathsf{DAT} \prec (\mathsf{WG}, \mathsf{CQ})$

Proof. $\mathsf{DAT} \preceq (\mathsf{WG}, \mathsf{CQ})$ holds trivially since a set of Datalog rules is a weakly-guarded set of tgds. In particular, a Datalog query $(\Sigma, \mathtt{Ans}/n)$ over $\mathbf{S}$ is equivalent to the query $(\mathbf{S}, \Sigma, \mathtt{Ans}(x_1, \ldots, x_n))$, where Σ is trivially weakly-guarded since there are no existentially quantified variables, which in turn implies that the set of affected positions of $sch(\Sigma)$ is empty. It remains to show that $(\mathsf{WG}, \mathsf{CQ}) \npreceq \mathsf{DAT}$. To this end, we employ a complexity-theoretic argument. It is well-known that the (decision version of the) problem of evaluating a Datalog query is feasible in polynomial time in data complexity, while for $(\mathsf{WG}, \mathsf{CQ})$ is complete for EXPTIME [7]. Thus, $(\mathsf{WG}, \mathsf{CQ}) \preceq \mathsf{DAT}$ implies that PTIME = EXP-TIME, which is a contradiction.   □

We finally show that there is no impact on the expressiveness of the query languages that are based on weakly-(frontier-)guarded sets of tgds:

Lemma 5. $(\mathsf{WG}, \mathsf{CQ}) = (\mathsf{WFG}, \mathsf{CQ}) = (\mathsf{WG}, \mathsf{CQ})_{\times} = (\mathsf{WFG}, \mathsf{CQ})_{\times}$.

Proof. It is well-known that $(\mathsf{WG}, \mathsf{CQ}) = (\mathsf{WFG}, \mathsf{CQ})$; $(\mathsf{WG}, \mathsf{CQ}) \preceq (\mathsf{WFG}, \mathsf{CQ})$ holds trivially since $\mathsf{WG} \subseteq \mathsf{WFG}$, while $(\mathsf{WFG}, \mathsf{CQ}) \preceq (\mathsf{WG}, \mathsf{CQ})$ has been shown in [9]. It remains to show $(\mathsf{WG}, \mathsf{CQ}) = (\mathsf{WG}, \mathsf{CQ})_{\times}$ and $(\mathsf{WFG}, \mathsf{CQ}) = (\mathsf{WFG}, \mathsf{CQ})_{\times}$. The $(\preceq)$ direction is trivial. The other direction holds since WG and WFG have the power to explicitly define a predicate P^k/k, where $k > 0$, that holds all the k-tuples of constants in the active domain. More precisely, a $(\mathsf{WG}, \mathsf{CQ})_{\times}$ (resp., $(\mathsf{WFG}, \mathsf{CQ})_{\times}$) query $Q = (\mathbf{S}, \Sigma, q)$, with $\alpha = \{j \mid \mathsf{Dom}^j \in \mathbf{S}\}$, is equivalent to the $(\mathsf{WG}, \mathsf{CQ})$ (resp., $(\mathsf{WFG}, \mathsf{CQ})$) query $Q' = (\mathbf{S}', \Sigma', q')$, where $\mathbf{S}' = \mathbf{S} \setminus \{\mathsf{Dom}^k \mid k \in \alpha\}$, Σ' is obtained from Σ by replacing each predicate Dom^k with P^k and adding the set of tgds:

$$R(x_1, \ldots, x_n) \to P^1(x_1), \ldots, P^1(x_n), \text{ for each } R \in \mathbf{S}'$$
$$P^1(x_1), \ldots, P^1(x_k) \to P^k(x_1, \ldots, x_k), \text{ for each } k \in \alpha,$$

and finally q' is obtained from q by replacing each predicate Dom^k with P^k.   □

It is now easy to verify that Lemmas 1, 2, 3, 4 and 5 imply Theorem 1.

6   Complexity of Query Evaluation

The question that remains to be answered is whether product databases have an impact on the complexity of the query evaluation problem under the guarded-based ontology-mediated query languages in question. As is customary when studying the computational complexity of the evaluation problem for a query language, we consider its associated decision problem. We denote this problem by EVAL($\mathcal{Q}$), where $\mathcal{Q}$ is an ontology-mediated query language, and its definition follows:

INPUT :	Query $Q = (\mathbf{S}, \Sigma, q(\bar{x})) \in \mathcal{Q}$, $\mathbf{S}$-database D, and tuple $\bar{t} \in \mathbf{C}^{|\bar{x}|}$.
QUESTION :	Does $\bar{t} \in cert_{q, \Sigma}(D)$?

It is important to say that when we focus on ontology-mediated queries over a product database, then the input database to the evaluation problem is product. In other words, if we focus on the problem EVAL$((\mathcal{C}, \mathsf{CQ})_{\times})$, where $\mathcal{C}$ is

Table 1. Complexity of EVAL$((\mathcal{C}, \mathsf{CQ})_\times)$; all the results are completeness results.

Class $\mathcal{C}$	Data complexity	Bounded arity	Combined complexity
G	PTIME	EXPTIME	2EXPTIME
FG	PTIME	2EXPTIME	2EXPTIME
WG	EXPTIME	EXPTIME	2EXPTIME
WFG	EXPTIME	2EXPTIME	2EXPTIME

a class of tgds, and the input query is $(\mathbf{S}, \Sigma, q)$, then the input database is an α-product database, where $\alpha = \{i \mid \mathsf{Dom}^i \in \mathbf{S}\}$. The complexity of EVAL$((\mathcal{C}, \mathsf{CQ}))$, where $\mathcal{C} \in \{\mathsf{G}, \mathsf{FG}, \mathsf{WG}, \mathsf{WFG}\}$, is well-understood; for $(\mathsf{G}, \mathsf{CQ})$ and $(\mathsf{WG}, \mathsf{CQ})$ it has been investigated in [7], while for $(\mathsf{FG}, \mathsf{CQ})$ and $(\mathsf{WFG}, \mathsf{CQ})$ in [3]. It is clear that the algorithms devised in [3,7] for the guarded-based ontology-mediated query languages in question treat product databases in the same way as non-product databases, or, in other words, they are oblivious to the fact that an input database is product. Therefore, we can conclude that, even if we focus on product databases, the existing algorithms can be applied and get the same complexity results for query evaluation as in the case where we consider arbitrary (non-product) databases; these results are summarized in Table 1. Recall that the data complexity is calculated by considering only the database as part of the input, while in the combined complexity both the query and the database are part of the input. We also consider the important case where the arity of the schema is bounded by an integer constant.

6.1 The Bounded Arity Case Revisited

In Table 1, the bounded arity column refers to the case where all predicates in the given query, including the predicates of the form Dom^k, where $k > 0$, are of bounded arity. However, bounding the arity of the Dom^k predicates is not our intention. Observe that in the proof of Lemma 3, where we show $(\mathsf{G}, \mathsf{CQ})_\times = (\mathsf{FG}, \mathsf{CQ})_\times = \mathsf{DAT}$, the predicates of the form Dom^k are used (i) to convert a CQ into a frontier-guarded tgd, and (ii) to convert a Datalog rule into a guarded tgd. More precisely, in the first case we use a Dom^k atom to guard the answer variables of a CQ, while in the second case to guard the variables in the body of a Datalog rule. Therefore, in both cases, we need to guard via a Dom^k atom an unbounded number of variables, even if the arity of the schema is bounded, and thus k must be unbounded. From the above discussion, it is clear that the interesting case to consider in our complexity analysis is not when all predicates of the underlying schema are of bounded arity, but when all predicates except the domain predicates are of bounded arity. Clearly, in case of $(\mathsf{FG}, \mathsf{CQ})_\times$ and $(\mathsf{WFG}, \mathsf{CQ})_\times$, the complexity of query evaluation is 2EXPTIME-complete, since the problem is 2EXPTIME-hard even if all predicates (including the domain predicates) have bounded arity. However, the picture is foggy in the case of $(\mathsf{G}, \mathsf{CQ})_\times$ and $(\mathsf{WG}, \mathsf{CQ})_\times$ since the existing results imply a 2EXPTIME upper bound and an EXPTIME lower bound. Interestingly, as we discuss below, the complexity of query evaluation remains the same, i.e., EXPTIME-complete, even if the domain predicates have unbounded arity.

Theorem 2. EVAL$((\mathsf{WG}, \mathsf{CQ})_\times)$ *is* EXPTIME-*complete if the arity of the schema, excluding the predicates of the form* Dom^k*, for* $k > 0$*, is bounded by an integer constant.*

The lower bound follows from the fact EVAL$((\mathsf{WG}, \mathsf{CQ}))$ is EXPTIME-hard when the arity of the schema is bounded [7]. The upper bound relies on a result that, although is implicit in [7], it has not been explicitly stated before. The *body-predicates* of an ontology-mediated query $(\mathbf{S}, \Sigma, q)$ are the predicates that do not appear in the head of a tgd of Σ. It holds that:

Proposition 1. EVAL$((\mathsf{WG}, \mathsf{CQ})_\times)$ *is in* EXPTIME *if the arity of the schema, excluding the body-predicates, is bounded by an integer constant.*

The above result simply states that even if we allow the body-predicates to have unbounded arity, while all the other predicates of the schema are of bounded arity, the complexity of EVAL$((\mathsf{WG}, \mathsf{CQ})_\times)$ remains the same as in the case where all the predicates of the schema have bounded arity. Since the predicates of the form Dom^k, for $k > 0$, is a subset of the body-predicates of an ontology-mediated query over a product database, it is clear that Proposition 1 implies Theorem 2. As said, although Proposition 1 has not been explicitly stated before, it is implicit in [7], where the complexity of query evaluation for $(\mathsf{WG}, \mathsf{CQ})$ is investigated. In fact, we can apply the alternating algorithm devised in [7] for showing that EVAL$((\mathsf{WG}, \mathsf{CQ}))$ is in EXPTIME if the arity of the schema (including the body-predicates) is bounded by an integer constant. In what follows, we briefly recall the main ingredients of the alternating algorithm proposed in [7], and discuss how we get the desired upper bound.

Recall that a set $\Sigma \in \mathsf{WG}$ can be effectively transformed into a set $\Sigma' \in \mathsf{WG}$ such that all the tgds of Σ' are single-head [7]. Henceforth, for technical clarity, we focus on tgds with just one atom in the head. Let D be a database, and Σ a set of tgds. Fix a chase sequence $D = I_0 \xrightarrow{\sigma_0, h_0} I_1 \xrightarrow{\sigma_1, h_1} I_2 \ldots$ for D under Σ. The instance $chase(D, \Sigma)$ can be naturally represented as a labeled directed graph $G = (N, E, \lambda)$ as follows: (1) for each atom $R(\bar{t}) \in chase(D, \Sigma)$, there exists $v \in N$ such that $\lambda(v) = R(\bar{t})$; (2) for each $i \geq 0$, with $I_i \xrightarrow{\sigma_i, h_i} I_{i+1}$, and for each atom $R(\bar{t}) \in h_i(body(\sigma_i))$, there exists $(v, u) \in E$ such that $\lambda(v) = R(\bar{t})$ and $\{\lambda(u)\} = I_{i+1} \setminus I_i$; and (3) there are no other nodes and edges in G. The *guarded chase forest* of D and Σ, denoted $\mathsf{gcf}(D, \Sigma)$, is the forest obtained from G by keeping only the nodes associated with weak-guards, and their children; for more details, we refer the reader to [7].

Consider a query $(\mathbf{S}, \Sigma, q) \in (\mathsf{WG}, \mathsf{CQ})$, a database D over $\mathbf{S}$, and a tuple $\bar{t}$ of constants. Clearly, $\bar{t} \in cert_{q,\Sigma}(D)$ iff there exists a homomorphism that maps $q(\bar{t})$ to $\mathsf{gcf}(D, \Sigma)$. Observe that if such a homomorphism h exists, then in $\mathsf{gcf}(D, \Sigma)$ there exist paths starting from nodes labeled with database atoms and ending at nodes labeled with atom of $h(q(\bar{t}))$. The alternating algorithm in [7] first guesses the homomorphism h from $q(\bar{t})$ to $\mathsf{gcf}(D, \Sigma)$, and then constructs in parallel universal computations the paths from D to $h(q(\bar{t}))$ (if they exist). During this alternating process, the algorithm exploits a key result established in [7], that is, the subtree of $\mathsf{gcf}(D, \Sigma)$ rooted at some atom $R(\bar{u})$ is determined by the so-called cloud of $R(\bar{u})$ (modulo renaming of nulls) [7, Theorem 5.16]. The cloud of $R(\bar{u})$ with respect to D and Σ, denoted $cloud(R(\bar{u}), D, \Sigma)$, is defined as $\{S(\bar{v}) \in chase(D, \Sigma) \mid \bar{v} \subseteq (adom(D) \cup \bar{u})\}$, i.e., the atoms in the result of the chase with constants from D and terms from $\bar{u}$. This result allows the algorithm to build the relevant paths of $\mathsf{gcf}(D, \Sigma)$ from D to $h(q(\bar{t}))$. Roughly, an atom $R(\bar{u})$ on a path can be generated by considering only its parent atom $S(\bar{v})$ and the cloud of $S(\bar{v})$ with respect to D and Σ. Whenever a new atom is generated, the algorithm nondeterministically guesses its cloud, and verify in a parallel universal computation that indeed belongs to the result of the chase.

From the above informal description, we conclude that the space needed at each step of the computation of the alternating algorithm is actually the size of the cloud of an atom. By applying a simple combinatorial argument, it is easy to show that the size of a cloud is at most $(|\mathbf{S}| + |sch(\Sigma)|) \cdot (|adom(D)| + w)^w$, where w is the maximum arity over all predicates of $\mathbf{S} \cup sch(\Sigma)$. Therefore, if we assume that all the predicates of the schema have bounded arity, which means that w is a constant, then the size of a cloud is polynomial. Since alternating polynomial space coincides with deterministic exponential time, we immediately get the EXPTIME upper bound in the case of bounded arity. Now, let $\mathbf{B}$ be the body-predicates of $(\mathbf{S}, \Sigma, q)$. It is clear that, for every atom $R(\bar{u}) \in chase(D, \Sigma)$, the restriction of $cloud(R(\bar{u}), D, \Sigma)$ on the predicates of $\mathbf{B}$ is actually D, and thus of polynomial size, even if the predicates of $\mathbf{B}$ have unbounded arity. This implies that even if we allow body-predicates of unbounded arity the size of a cloud remains polynomial. Therefore, the alternating algorithm devised in [7] can be applied in order to get the EXPTIME upper bound stated in Proposition 1.

7 Conclusions

It is realistic to assume that a database management system provides access to the active domain via built-in relations, or, in more formal terms, to assume that queries are evaluated over product databases. Interestingly, the query languages that are based on (frontier-)guarded existential rules gain in expressiveness when we focus on product databases; in fact, they have the same expressive power as Datalog. On the other hand, there is no impact on the expressive power of the query languages based on weakly-(frontier-)guarded existential rules, since they are powerful enough to explicitly compute the predicates needed to access the active domain. We also observe that there is no impact on the computational complexity of the query languages in question.

Acknowledgements. Gottlob is supported by the EPSRC Programme Grant EP/M025268/. Pieris and Šimkus are supported by the Austrian Science Fund (FWF), projects P25207-N23 and Y698.

References

1. Abiteboul, S., Hull, R., Vianu, V.: Foundations of Databases. Addison-Wesley, Boston (1995)
2. Baget, J.F., Leclère, M., Mugnier, M.L., Salvat, E.: On rules with existential variables: walking the decidability line. Artif. Intell. **175**(9–10), 1620–1654 (2011)
3. Baget, J.F., Mugnier, M.L., Rudolph, S., Thomazo, M.: Walking the complexity lines for generalized guarded existential rules. In: IJCAI, pp. 712–717 (2011)
4. Bárány, V., Benedikt, M., ten Cate, B.: Rewriting guarded negation queries. In: Chatterjee, K., Sgall, J. (eds.) MFCS 2013. LNCS, vol. 8087, pp. 98–110. Springer, Heidelberg (2013)
5. Beeri, C., Vardi, M.Y.: The implication problem for data dependencies. In: Even, S., Kariv, O. (eds.) ICALP 1981. LNCS, vol. 115, pp. 73–85. Springer, Heidelberg (1981)
6. Bienvenu, M., ten Cate, B., Lutz, C., Wolter, F.: Ontology-based data access: a study through disjunctive datalog, csp, and MMSNP. ACM Trans. Database Syst. **39**(4), 33:1–33:44 (2014)
7. Calì, A., Gottlob, G., Kifer, M.: Taming the infinite chase: query answering under expressive relational constraints. J. Artif. Intell. Res. **48**, 115–174 (2013)
8. Fagin, R., Kolaitis, P.G., Miller, R.J., Popa, L.: Data exchange: semantics and query answering. Theor. Comput. Sci. **336**(1), 89–124 (2005)
9. Gottlob, G., Rudolph, S., Simkus, M.: Expressiveness of guarded existential rule languages. In: PODS, pp. 27–38 (2014)
10. Johnson, D.S., Klug, A.C.: Testing containment of conjunctive queries under functional and inclusion dependencies. J. Comput. Syst. Sci. **28**(1), 167–189 (1984)
11. Maier, D., Mendelzon, A.O., Sagiv, Y.: Testing implications of data dependencies. ACM Trans. Database Syst. **4**(4), 455–469 (1979)

Negative Knowledge for Certain Query Answers

Leonid Libkin[(✉)]

School of Informatics, University of Edinburgh, Edinburgh, UK
libkin@inf.ed.ac.uk

Abstract. Querying incomplete data usually amounts to finding answers we are certain about. Standard approaches concentrate on positive information about query answers, and miss negative knowledge, which can be useful for two reasons. First, sometimes it is the only type of knowledge one can infer with certainty, and second, it may help one find good and efficient approximations of positive certain answers. Our goal is to consider a framework for defining both positive and negative certain knowledge about query answers and to show two applications of it. First, we demonstrate that it naturally leads to a way of representing certain information that has hitherto not been used in querying incomplete databases. Second, we show that approximations of such certain information can be computed efficiently for all first-order queries over relational databases.

1 Introduction

If uncertainty occurs in a dataset, answering queries against it typically involves computing *certain answers*, i.e., answers one can be sure about. This happens in traditional database query answering [2,22] and in numerous applications such as data integration [26], data exchange [4], inconsistent databases [7], and ontology-based data access [11,25]. The most common approach is to look at all complete datasets D' that can potentially represent an incomplete dataset D – i.e., its *semantics* $[\![D]\!]$ – and answers that are true in all such D'. When a query Q returns sets of objects (for example, sets of tuples for relational database queries), certainty is typically defined by $\mathsf{certain}(Q, D) = \bigcap\{Q(D') \mid D' \in [\![D]\!]\}$, see [30]. This definition has been so dominant in the literature that even in models where queries do not return sets, languages have been adjusted to make this definition applicable (e.g., for XML and graph data [3,5,6]).

Certain answers defined this way can be viewed as a variant of the logical validity problem. This, not surprisingly, leads to high complexity bounds; in fact, query answering tends to be tractable for conjunctive queries or relatives, but computationally infeasible beyond [1,4,5,7,8,10,26,35]. A very common situation is that adding features to conjunctive queries or their unions makes finding certain answers coNP-hard or even undecidable. It is thus well understood that the inability of the standard theoretical solutions to handle the problem of querying incomplete data outside a limited class of queries needs to be addressed. Recently, two lines of work in this direction have been pursued. The first revisits

M. Ortiz and S. Schlobach (Eds.): RR 2016, LNCS 9898, pp. 111–127, 2016.
DOI: 10.1007/978-3-319-45276-0_9

the very notion of certainty in query answering, and the second attempts to approximate certain answers efficiently.

The first line of work in fact dates back to the 1980s, when an alternative (and, as several papers [24,28] have argued, better) definition of certain answers appeared [31]. More recently, a general and data model-independent approach to defining query answers over incomplete databases was proposed in [28]. It was based on combining classical data management techniques with viewing databases as logical theories, as advocated by [33,34], as well as using the idea of ordering incomplete databases in terms of their informativeness [9]. Certain answers can be represented by logical formulae true about answers in all possible worlds, and the notion of certainty is closely connected to logical entailment, rather than an arbitrary choice of intersection in the definition of certain. Using informativeness ordering, one can state when a query answering algorithm behaves rationally: this happens if it produces more informative answers on more informative inputs. For relational databases, these ideas led to new large classes of queries for which certain answers can be computed efficiently [17], and to a new account of many-valued query answers [13], as employed by all standard DBMSs [14].

The second line of work, based on approximations, was also used recently to show that an efficient approximation of certain answers can be computed for all first-order queries [29], not just unions of conjunctive queries, as was previously known [22]. A crucial element of that approach is that one needs to carry *negative certain* information while computing the answer, although at the end such negative information is dismissed and only the positive answer is given to the user.

However, dismissing negative information is not always a good path to follow, as it may in fact provide us with useful information about query answers. For example, consider a database D with two unary relations R and S, so that R contains an unknown value (a *null* in the database terminology), $S = \{1\}$, and the query $Q(x) = R(x) \wedge \neg S(x)$ computes their difference. Then the certain answer is empty under every reasonable semantics. But we can be certain that 1 is not in the answer; hence, we are certain about the fact $\neg A(1)$ (with A for "answer"), which says that while we do not know what may occur in the output, we do know that 1 does not occur. Even though in this example $A(1)$ is the certain answer for the negation of Q, in general certain negative answers to Q are not the same as what is known with certainty about $\neg Q$. Indeed, consider relations $R' = \{(1, \perp)\}$ and $S' = \{(1, \perp')\}$, where $\perp, \perp'$ indicate nulls (not necessarily denoting the same value). The negation of $Q'(\bar{x}) = R'(\bar{x}) \wedge \neg S'(\bar{x})$ is $S'(\bar{x}) \vee \neg R'(\bar{x})$, and thus, with certainty, the answer to $\neg Q'$ will have a tuple whose first component is 1, i.e., we know $\exists y A(1, y)$ about the answer to Q'. However, we cannot tell which tuples with certainty do *not* belong to the answer to Q'.

Even these simple examples tell us that the user may benefit from having negative certain information about query answers, and getting it involves more than just finding certain answers for the negation of the query. To understand how such negative information can be incorporated into query answering, we need to address several questions:

(a) How do we define negative and positive knowledge about query answers, and what is the connection between the two?

(b) In what logical languages can we express such negative knowledge? Can such knowledge be represented in a user-friendly way, and if so, does it correspond to any of the known ways of defining query answers?

(c) What is the complexity of finding positive and negative knowledge about query answers?

(d) If exact computation is infeasible, can we effectively approximate answers with some guarantees?

To answer the first question, we follow the approach of [17, 28] which treats incompleteness at an abstract level applicable to many data models. The key elements of the approach are the notions of complete and incomplete models, the *semantics* of an incomplete object, which is a set of complete ones it can denote, and a set of formulae representing *knowledge* about objects. The semantic function makes it possible to define *informativeness ordering*, which says when one object is more informative than another. The restriction of frameworks in [17, 28] was that knowledge was *positive*: if a fact is known about an object, it remains true in more informative ones. Negative knowledge is not such: we can think of it as saying that we do not know some fact about an object; therefore, we do not know that fact about less informative objects.

Positive formulae were used in [28] to define certain knowledge about sets of objects, providing a disciplined notion of certain answers, rather than an ad hoc one based on the notion of intersection. The idea is as follows: the *theory* of a set of objects is everything we know about that set with certainty. Such a theory of course could be infinite, but if we find a single formula equivalent to it, then this formula gives us a proper representation of certain knowledge.

To see what kinds of formulae we can use for negative knowledge, we follow a similar approach, but conditions required for good behavior of negative knowledge impose significant computational requirements, despite a seemingly simple reversal of the ordering. But we turn this to our advantage and use such conditions as a guide for finding logical formalisms for negative formulae. For relational databases, this results in a new formalism that exhibits a *duality* between formulae and objects, making it possible to apply effective query evaluation to compute certain knowledge.

This new formalism for defining certain answers (both positive and negative) is closely related to standard approaches used in the literature [22, 31] and yet is not covered by them. In essence, it allows nulls from the input database to be present in query answers (which is more than [22] does) but only allows repetitions of such nulls within a single tuple (as opposed to [31], which allows repetitions across different tuples in the answer).

To demonstrate the usefulness of this approach and the new representation mechanism for relational databases, we show how to compute both positive and negative knowledge about certain answers for all first-order (equivalently, relational algebra/calculus) queries over relational databases. Given the intractability of certain answers even for Boolean first-order queries [1], our procedure gives an approximation for those, which is efficient, and comes with correctness guarantees.

Organization. Background material is presented in Sect. 2. Modeling negative knowledge is described in Sect. 3, and certain negative knowledge is studied in Sect. 4. Section 5 explains how to represent such knowledge for relational databases, and in Sect. 6 we provide an efficient algorithm for computing it.

2   Preliminaries

A General Model. We now recall the basic setting of [17,28] that lets us talk about the essential features of incompleteness without recourse to a particular data model. The two basic concepts are *objects*, and *formulae* they satisfy. Objects could be incomplete or complete; the semantics of an incomplete object is the set of complete objects it may represent.

Formally, a *database domain* is a triple $\mathbb{D} = \langle \mathcal{D}, \mathcal{C}, [\![\ ]\!] \rangle$, where $\mathcal{D}$ is a set of objects (for instance, all relational databases over the same schema), $\mathcal{C}$ is the set of complete objects (for instance, databases over the same schema without incomplete information), and $[\![\ ]\!] : \mathcal{D} \to 2^{\mathcal{C}}$ is the semantic function: $[\![x]\!] \subseteq \mathcal{C}$ is the semantics of an object x. We require that a complete object denote at least itself: if $c \in \mathcal{C}$, then $c \in [\![c]\!]$.

The *information ordering* is defined by

$$x \preceq y \;\Leftrightarrow\; [\![y]\!] \subseteq [\![x]\!]. \tag{1}$$

That is, the more an object denotes, the less we know about it (indeed, if we know nothing about something, it can denote everything). We require that objects in the semantics of x be at least as informative as x: if $c \in [\![x]\!]$, then $x \preceq c$. This condition holds for all the standard semantics of incompleteness.

We also assume that we have a set of formulae $\mathbb{F}$ that express knowledge about objects in $\mathcal{D}$ and a satisfaction relation $\models$ between $\mathcal{D}$ and $\mathbb{F}$; that is, $x \models \varphi$ if φ is true in x. For sets of objects and formulae, we write $X \models \varphi$ if $x \models \varphi$ for each $x \in X$, and $x \models \Phi$ if $x \models \varphi$ for each $\varphi \in \Phi$. As usual, $\mathsf{Th}(X) = \{\varphi \mid X \models \varphi\}$ is the *theory* of X, and $\mathsf{Mod}(\Phi) = \{x \mid x \models \Phi\}$ is the set of models of Φ.

Previously, only *positive* knowledge was considered, i.e., it was required that $x \preceq y$ and $x \models \varphi$ imply $y \models \varphi$.

For domains $\mathbb{D} = \langle \mathcal{D}, \mathcal{C}, [\![\ ]\!] \rangle$ and $\mathbb{D}' = \langle \mathcal{D}', \mathcal{C}', [\![\ ]\!]' \rangle$, a *query* is modeled as a mapping $Q : \mathcal{D} \to \mathcal{D}'$ such that $Q(c) \in \mathcal{C}'$ whenever $c \in \mathcal{C}$ (no incompleteness is introduced when a query acts on a complete object). Note that the semantics $[\![\ ]\!]'$ of query *answers* need not be the same as the semantics $[\![\ ]\!]$ of query inputs.

The main object one then works with [2,22] is

$$Q([\![x]\!]) \;=\; \{Q(c) \mid c \in [\![x]\!]\} \subseteq \mathcal{D}' \tag{2}$$

which gives us the answers to Q in all possible worlds representing x. Finding certain answers to Q on x then amounts to extracting what we know with certainty about $Q([\![x]\!])$.

Certain Knowledge. Since computing certain answers amounts to extracting certain information from a set of objects, typically of the form (2), we need to know how to describe certain information in a set $X \subseteq \mathcal{D}$. We know that $\mathsf{Th}(X)$ is the set of facts that are true in all objects of X, i.e., this is what we know about X with certainty. The whole theory is not an object we want to work with for performing computational tasks (to start with, it is likely to be infinite). What we want instead is a single formula equivalent to this theory; then such a formula describes all the certain knowledge of X. Of course formulae/theories are equivalent when they have the same models. Using this, [28] proposed to define certain knowledge of a set of objects as a formula $\square X$ such that

$$\mathsf{Mod}(\square X) \;=\; \mathsf{Mod}(\mathsf{Th}(X)) \tag{3}$$

Such a formula may not exist for all sets X (by a simple cardinality argument), although in many cases relevant for query answering, it does. It need not be unique, but this is not a problem: if both $\mathsf{Mod}(\varphi_1)$ and $\mathsf{Mod}(\varphi_2)$ equal $\mathsf{Mod}(\mathsf{Th}(X))$, then φ_1 and φ_2 are equivalent, as formulae having the same models, and hence either one can be used as $\square X$.

Incomplete Relational Databases. As a concrete example of incomplete information, we consider relational databases with naïve, or marked nulls [2,22]. This model dominates in applications such as exchange and integration of data [4,26], and subsumes the usual model of nulls implemented in commercial DBMSs. In this model, there are two types of values: *constants* and *nulls*. There are countably infinite sets Const of constants (e.g., $1, 2, \ldots$), and Null of nulls, which will be denoted by $\bot$, with sub- or superscripts.

A relational *vocabulary* (or *schema*) is a set of relation names, each with its arity. An incomplete relational database D associates with each k-ary relation symbol R from the vocabulary a k-ary relation $R^D \subseteq (\mathsf{Const} \cup \mathsf{Null})^k$. When D is clear from the context, we write R rather than R^D. Sets of constants and nulls that occur in D are denoted by $\mathsf{Const}(D)$ and $\mathsf{Null}(D)$. The *active domain* of D is $\mathrm{adom}(D) = \mathsf{Const}(D) \cup \mathsf{Null}(D)$. A *complete* database D has no nulls, i.e., $\mathrm{adom}(D) \subseteq \mathsf{Const}$.

The basic semantics of incomplete databases is given by means of special kinds of homomorphisms between instances. A map $h : \mathsf{Null} \to \mathsf{Const} \cup \mathsf{Null}$ is a *homomorphism* between two instances D and D' if for each relation symbol R, if $\bar{t} \in R^D$, then $h(\bar{t}) \in R^{D'}$. Here $h(v_1, \ldots, v_k) = (h(v_1), \ldots, h(v_k))$, and we assume that $h(v) = v$ for each $a \in \mathsf{Const}$.

A homomorphism is called a *valuation* if $h(v) \in \mathsf{Const}$ for each v. By $h(D)$ we denote the image of a homomorphism, i.e., the database consisting of all the tuples $h(\bar{t})$ for $\bar{t} \in R^D$, for each relation R in the vocabulary.

The standard semantics of incompleteness [22] are the *closed world assumption* (CWA) and the *open world assumption* (OWA) semantics:

$$[\![D]\!]_{\mathrm{CWA}} \;=\; \{h(D) \mid h \text{ is a valuation}\},$$

$$[\![D]\!]_{\mathrm{OWA}} \;=\; \{h(D) \cup D' \mid h \text{ is a valuation}, \ D' \text{ is complete}\}.$$

The former simply replaces nulls by constants, and the latter in addition allows us to add any set of complete tuples.

The information orderings (1) given by these semantics are as follows: for OWA, $D \preceq_{OWA} D'$ iff there is a homomorphism $h : D \to D'$, and for CWA, $D \preceq_{CWA} D'$ iff there is a homomorphism $h : D \to D'$ such that $D' = h(D)$, see [17].

Queries. A *relational query* of arity k maps databases D over a relational schema into a single k-ary relation, which we denote here by A (for 'answers'). This is in line with standard languages such as relational calculus, relational algebra, and SQL, whose queries specify attributes of an output table [2,14].

The classical definition [22] of certain answers in the literature is the set $\mathrm{certain}(Q, D)$ of tuples $\bar{u}$ over Const such that $\bar{u} \in Q(D')$ for every $D' \in [\![D]\!]$. Note that answers depend on the semantics of the input. Another definition, which has the advantage of keeping nulls in answers, is that of *certain answers with nulls*, $\mathrm{certain}_\perp(Q, D)$ (it was first defined in [31] although not given a name; the name we use is from [29]). For CWA, the set $\mathrm{certain}_\perp(Q, D)$ consists of all tuples $\bar{u}$ over $\mathrm{adom}(D)$ – thus having both constants and nulls – such that for every valuation h on D, we have $h(\bar{u}) \in Q(h(D))$. It turns out that, under CWA, $\mathrm{certain}(Q, D)$ is precisely the set of constant tuples in $\mathrm{certain}_\perp(Q, D)$.

For relational databases, as our basic language we consider *first-order logic* (FO) over the relational vocabulary (i.e., relational calculus, which also serves as the basis of SQL [2]). More precisely, its atomic formulae are relational atoms $R(\bar{x})$ and equality atoms $x = y$, and its formulae are closed under Boolean connectives $\wedge, \vee, \neg$ and quantifiers $\exists, \forall$. The $\exists, \wedge$-closure of atomic formulae is referred to as the set of *conjunctive queries*; those are of the form $\varphi(\bar{x}) = \exists \bar{z} \bigwedge_i R_i(\bar{z}_i)$ where each R_i is a relation symbol and variables in tuple $\bar{z}_i$s come from $\bar{x}$ and $\bar{y}$.

3   Modeling Negative Knowledge

So far we assumed that the knowledge of objects is positive: a formula φ true in an object continues to be true when an object is replaced by a more informative one. While in general we often deal with logical formalisms not closed under negation (e.g., conjunctive queries), assume for now that we can negate φ. If $\neg\varphi$ is true an object x, and $y \preceq x$, then we would have $y \models \neg\varphi$. Thus, to model negative knowledge in general, we look at formulae whose sets of models are downward closed. In other words, we now have two sets of formulae, $\mathbb{F}^+$ and $\mathbb{F}^-$, such that,

- for $\varphi \in \mathbb{F}^+$, if $x \preceq y$ and $x \models \varphi$, then $y \models \varphi$;
- for $\psi \in \mathbb{F}^-$, if $x \preceq y$ and $y \models \psi$, then $x \models \psi$.

There appear to be two possible approaches to extending the framework of [28] with both certain positive and certain negative knowledge.

The first approach. We follow the idea behind the definition (3). We can define $\mathsf{Th}^+(X) = \{\varphi \in \mathbb{F}^+ \mid X \models \varphi\}$ and $\mathsf{Th}^-(X) = \{\psi \in \mathbb{F}^- \mid X \models \psi\}$ as theories expressing positive and negative knowledge about X, and then, as in (3), try to capture them with formulae $\square^+ X$ and $\square^- X$ such that

$$\mathsf{Mod}(\square^+ X) = \mathsf{Mod}(\mathsf{Th}^+(X))$$
$$\mathsf{Mod}(\square^- X) = \mathsf{Mod}(\mathsf{Th}^-(X)) \tag{4}$$

When they exist, these formulae represent certain positive knowledge and certain negative knowledge about X. Note that $\mathsf{Mod}(\square^+ X)$ is upward-closed and $\mathsf{Mod}(\square^- X)$ is downward-closed with respect to $\preceq$.

The second approach. Note that (3) is based on an equivalence between two theories: $\Phi \approx_{\mathsf{tt}} \Psi$ whenever for each object x, all formulae of Φ are true in x iff all formulae of Φ are true in x. Then we just required that $\square^+ X \approx_{\mathsf{tt}} \mathsf{Th}^+(X)$.

An alternative is to look at equivalence with respect to negative information, essentially changing true and false. We let $\Phi \approx_{\mathsf{ff}} \Psi$ whenever for each object x, all formulae of Φ are false in x iff all formulae of Ψ are false in x. It would make sense then to capture all things we know to be false in X using this equivalence. That is, we define

$$\mathsf{Th}^+_\neg(X) = \{\varphi \in \mathbb{F}^+ \mid X \models \neg\varphi\}$$
$$\mathsf{Th}^-_\neg(X) = \{\psi \in \mathbb{F}^- \mid X \models \neg\psi\}$$

as sets of formulae we know with certainty are false in X, and then try to capture them with single formulae satisfying

$$\square^+_\neg X \approx_{\mathsf{ff}} \mathsf{Th}^+_\neg(X) \text{ and } \square^-_\neg X \approx_{\mathsf{ff}} \mathsf{Th}^-_\neg(X). \tag{5}$$

Both of these seem to be reasonable ways of capturing negative information about a set of objects; fortunately, they are closely related so we can choose either (4) or (5) as the main definition. For formulae α, β, we write $\alpha = \neg\beta$ if $\mathsf{Mod}(\alpha) = \mathcal{D} - \mathsf{Mod}(\beta)$ (so $\alpha = \neg\beta$ implies $\beta = \neg\alpha$).

Theorem 1. *Assume that $\mathbb{F}^-$ contains exactly the negations of formulae in $\mathbb{F}^+$. If formulae $\square^* X$ and $\square^*_\neg X$ exist, when $*$ is $+$ or $-$, we have the following relationships between them:* $\square^+ X = \neg\square^-_\neg X$ *and* $\square^- X = \neg\square^+_\neg X$.

To illustrate the difference between two ways of representing negative information, consider a database with relations $R = \{\bot\}$ and $S = \{1,2\}$, and a query Q that computes their difference $R - S$, i.e., $Q(x) = R(x) \wedge \neg S(x)$. Let $X = Q(\llbracket R, S \rrbracket)$ under either OWA or CWA, and consider $\mathbb{F}^+$ that consists of atomic relational formulae. Then $\mathsf{Th}^+_\neg(X)$ contains $A(1)$ and $A(2)$, and thus $\square^+_\neg X$ is equivalent to $A(1) \vee A(2)$. That is, $\square^+_\neg X$ describes what we know with certainty will *not* hold in the answer to the query. On the other hand, $\square^- X$ is equivalent to $\neg A(1) \wedge \neg A(2)$ (again, assuming the connection between $\mathbb{F}^+$ and $\mathbb{F}^-$ as in the theorem) and describes negative information that is guaranteed to be true in the query result.

Certain Knowledge for Query Answering. Given an object x and a query Q, answering Q on x in a way that provides both positive and negative knowledge amounts to finding the pair of formulae

$$\Box(Q, x) \; = \; \left(\Box^+ Q([\![x]\!]), \, \Box^- Q([\![x]\!]) \right) \tag{6}$$

whenever such formulae exist, and their computation is feasible. For representing the second component, we can choose either $\Box^- Q([\![x]\!]))$, or its negation $\Box^+_- Q([\![x]\!])$, as Theorem 1 suggests. The components of (6) are the most general formulae defining positive and negative knowledge, as they imply all formulae in $\mathsf{Th}^+(Q([\![x]\!]))$ and $\mathsf{Th}^-(Q([\![x]\!]))$, respectively. If computing them is infeasible, we can look for *approximations* by means of returning a pair (α, β) of formulae such that $\alpha \in \mathsf{Th}^+(Q([\![x]\!]))$ and $\beta \in \mathsf{Th}^-(Q([\![x]\!]))$. They may not be as general as (6), but they still give us information about query answers we can be certain about.

4   Certain Negative Knowledge

Our next goal is to understand how to represent certain negative and positive information, particularly for sets X which are possible query answers, as in (2). That is, we will see what requirements must be imposed on logical formalisms $\mathbb{F}^+$ and $\mathbb{F}^-$ to ensure feasible computation of certain answers.

Towards understanding these requirements, we present an alternative view of formulae $\Box^+ X$ and $\Box^- X$. For that, consider the usual implication of formulae, $\varphi \supset \psi$ iff $\mathsf{Mod}(\varphi) \subseteq \mathsf{Mod}(\psi)$. It generates a preorder (reflexive transitive relation) on sets $\mathbb{F}^+$ and $\mathbb{F}^-$. Viewing implication as a preorder, we define, for a set of formulae Φ, the formula $\bigwedge \Phi$ as the *greatest lower bound* in the preorder $\supset$. That is, $\bigwedge \Phi \supset \Phi$ and whenever $\varphi' \supset \Phi$, we have that $\varphi' \supset \bigwedge \Phi$ (here $\varphi' \supset \Phi$ means that φ' implies every formula $\varphi \in \Phi$). These formulae let us capture certain knowledge provided by Φ, so it seems desirable to have $\Box^* X$ to be the same as $\bigwedge \mathsf{Th}^*(X)$, for $*$ being $+$ or $-$. We now explain when this is possible.

First, we remark that formulae $\bigwedge \Phi$ may not exist in general, and if they exist, they may not be unique, although any two such formulae are logically equivalent since they have the same models. While the notation $\bigwedge$ is standard for greatest lower bounds, the connection with conjunction is natural: if there is a formula φ equivalent to the conjunction of all formulae in Φ, then $\mathsf{Mod}(\varphi) = \mathsf{Mod}(\Phi)$ and $\varphi = \bigwedge \Phi$; in general though we may have $\mathsf{Mod}(\bigwedge \Phi) \subsetneq \mathsf{Mod}(\Phi)$.

We now show when $\Box^* X = \bigwedge \mathsf{Th}^*(X)$. In fact, for Th^+, this was already shown in [28], but under additional conditions that we now eliminate.

Let $\uparrow x = \{y \mid x \preceq y\}$ and $\downarrow x = \{y \mid y \preceq x\}$. By $\delta_x^\uparrow$ and $\delta_x^\downarrow$ we denote formulae (if they exist) such that $\mathsf{Mod}(\delta_x^\uparrow) = \uparrow x$ and $\mathsf{Mod}(\delta_x^\downarrow) = \downarrow x$.

Theorem 2. – If $\mathbb{F}^+$ is closed under conjunction and contains formulae $\delta_x^\uparrow$ for all x, then $\Box^+ X = \bigwedge \mathsf{Th}^+(X)$ for every X.
– If $\mathbb{F}^-$ is closed under disjunction and contains formulae $\delta_x^\downarrow$ for all x, then $\Box^- X = \bigwedge \mathsf{Th}^-(X)$ for every X.

The meaning of the equalities $\Box^* X = \bigwedge \mathsf{Th}^*(X)$ is that if one formula exists, then so does the other, and the two are equivalent, i.e., have the same models.

For most common semantics of incompleteness, formulae $\delta_x^\uparrow$ are easy to construct, and in fact they determine the shape of queries that can be answered easily under those semantics [17]. For instance, under OWA, they are conjunctive queries, and for CWA, they extend positive FO formulae with a limited form of guarded negation [12]. The new condition for Th^- that formulae $\delta_x^\downarrow$ be definable is harder to achieve, and this condition will let us choose the appropriate logical language for $\mathbb{F}^-$.

5 Representation of Relational Query Answers: Incomplete Tuples

We now use the abstract results of two previous sections to suggest a representation mechanism for relational query answers, and to show how to find positive and negative answers using such a representation. For finding a representation mechanism, we analyze computational properties of formulae $\delta_x^\uparrow$ and $\delta_x^\downarrow$. Restricting those to a tractable class, gives us a representation of answers, called *incomplete tuples*. This representation exhibits a *duality* between formulae and objects: that is, positive and negative theories of query answers can be viewed as set of conventional tuples that use null values. With this duality, we define query answers using (6). To check that the definition makes sense, we have to make sure that it preserves informativeness. This, in turn, means that we need to define orderings on query answers, i.e., sets of incomplete tuples. We do so, and then prove, in Theorem 3 that the resulting representation mechanism and query answering by means of (6) do behave rationally, i.e., preserve informativeness.

We start by looking at the requirements of Theorem 2 and analyzing formulae $\delta_x^\uparrow$ and $\delta_x^\downarrow$. While the former are easy to obtain for standard semantics of incompleteness, the latter could become too expensive computationally, and it is their complexity that will suggest the representation of positive and negative certain answers.

We deal with relational databases, as described in Sect. 2. When we deal with outputs of relational queries, which are sets of tuples, it suffices to deal with one predicate for each type of answers, positive or negative (of course usual relational databases just return one set of tuples, for positive answers). As before, we refer to that predicate as $A(\cdot)$; later, when we look in more detail at separation of positive and negative answers, we shall use predicates $A^+(\cdot)$ and $A^-(\cdot)$.

The first observation shows that one must impose rather strong restriction on the types of formulae $\mathbb{F}^+$ that represent query answers (note that this does *not* imply any restriction on queries themselves).

Proposition 1. *For the class of conjunctive queries, data complexity of formulae $\delta_A^\downarrow$ for relations A is in* NP; *in fact there is a relation for which data complexity of $\delta_A^\downarrow$ is* NP-complete.

Indeed, formulae $\delta_A^{\downarrow}$ test the existence of a homomorphism into A, i.e., they encode the general constraint satisfaction problem. In particular, such formulae are not expressible in FO, nor even its extensions with least and inflationary fixpoints.

Incomplete Tuples. The standard representation of query answers in relational databases is by means of ground tuples: one simply says that a tuple $\bar{a}$ is in the answer, or that predicate $A(\bar{a})$ holds. Proposition 1 says that extending it to conjunctive queries as a representation mechanism is too much from the complexity point of view. Over the vocabulary $A(\cdot)$ of query answers, Boolean conjunctive queries are of the form $\exists \bar{x}(A(\bar{x}_1, \bar{c}_1) \wedge \ldots \wedge A(\bar{x}_m, \bar{c}_m))$, where $\bar{c}_i$s are tuples of constants from Const and $\bar{x}_i$s are tuples of variables that together form $\bar{x}$. Eliminating variables gives us sets of constant tuples, i.e., the usual database query answers over complete data. Another way of simplifying the definition is to eliminate variables that occur in more than one $\bar{x}_i$, i.e., looking at formulae $\exists \bar{x}_1 A(\bar{x}_1, \bar{c}_1) \wedge \ldots \wedge \exists \bar{x}_m A(\bar{x}_m, \bar{c}_m)$. That is, we are dealing with conjunctions of formulae $\exists \bar{x} A(\bar{x}, \bar{c})$.

We can think of such formulae $\exists \bar{x} A(\bar{x}, \bar{c})$ as incomplete tuples. An *incomplete tuple* is simply a tuple of Const $\cup$ Null. There is a natural correspondence between formulae $\exists \bar{x} A(\bar{x}, \bar{c})$ and incomplete tuples: for instance, $\exists x, x' A(x, 1, x, 2, x')$ can be thought of as an incomplete tuple $(\perp, 1, \perp, 2, \perp')$. Note that this duality between incomplete tuples as formulae and as actual tuples lets us represent query answers of the form (6) just as database relations.

Representation of answers by means of incomplete tuples is between the usual marked nulls and the Codd interpretation of nulls, which model SQL's view of nulls [2,22]. Marked nulls can be repeated, and appear in different tuples; Codd nulls cannot be repeated at all. In incomplete tuples, a null can be repeated, but only within a tuple, and not across several tuples.

Query Answering and Ordering. We now consider orderings on query answers which are viewed as sets of incomplete tuples. Recall that we expect a rationally behaving query answering to produce more informative answers when more informative inputs are given; hence orderings are necessary to prove such rationality. For input databases, we have seen some standard orderings such as $\preceq_{\text{OWA}}$ and $\preceq_{\text{CWA}}$. According to (6), a query answer will be given as a pair of sets (A^+, A^-) of incomplete tuples. Tuples in A^+ belong to the answer with certainty; thus, when viewed as formulae, their conjunction is equivalent to the formula $\square^+ Q([\![x]\!])$. Tuples in A^- are those that certainly do not belong to the answer; hence conjunction of their negations is equivalent to $\square^- Q([\![x]\!])$.

First, we need to see how we can order incomplete tuples in terms of their informativeness. There are two ways of looking at it:

- What improves informativeness of a tuple? Replacing a null with a constant does, and replacing a null with another null might (e.g., if we replace $\perp'$ with $\perp$ in $(\perp, \perp')$, we get a more informative tuple $(\perp, \perp)$ giving extra information that its components are the same). Thus, given two incomplete tuples $\bar{a}$ and

$\bar{b}$ over Const $\cup$ Null, $\bar{b}$ is more informative than $\bar{a}$ if there is a homomorphism h so that $h(\bar{a}) = \bar{b}$.

- Alternatively, we view incomplete tuples as formulae and say that $\bar{b}$ is more informative than $\bar{a}$ if it logically entails it, i.e., $\bar{b} \supset \bar{a}$.

The homomorphism theorem for conjunctive queries tell us that these two are equivalent, so we can take either of them as the definition of $\bar{b}$ being more informative than $\bar{a}$, which will be denoted by $\bar{a} \lhd \bar{b}$.

Next, we look at sets of incomplete tuples A and B, and define orderings $\preceq^+_{\mathrm{IT}}$ and $\preceq^-_{\mathrm{IT}}$ saying that one of the sets has more positive or negative information than the other. First,

$$A \preceq^+_{\mathrm{IT}} B \quad \Leftrightarrow \quad \forall \bar{a} \in A \, \exists \bar{b} \in B : \ \bar{a} \lhd \bar{b}.$$

This ordering says that we can improve an answer by improving individual tuples in it, or adding new tuples that our initial attempt to approximate query answers may have missed. This is the ordering on positive query answers we shall use. Note that it is also consistent with observations made in [13,28] that for query answers (as opposed to inputs), the prefer interpretation is open-world, as adding tuples improves the answer.

When it comes to negative information, if we are given two incomplete tuples $\bar{a}$ and $\bar{b}$ such that $\bar{a} \lhd \bar{b}$, then it is actually better to have $\bar{a}$ in the answer, as it gives us more information about tuples to exclude. For instance, having a tuple $(1, 2)$ in the negative answer simply says that $(1, 2)$ is never in the answer, but having a tuple $(1, \bot)$ is more informative as it says that no tuple whose first component is 1 is in the answer. This leads to the following ordering:

$$A \preceq^-_{\mathrm{IT}} B \quad \Leftrightarrow \quad \forall \bar{b} \in B \, \exists \bar{a} \in A : \ \bar{a} \lhd \bar{b}.$$

Note that these are well-known orderings in the semantics of concurrency (so called Hoare and Smyth powerdomain orderings [20]) where they are used to compare possible outcomes of different threads of concurrent computations in terms of the information they carry. In terms of computational problems, unlike the relations $\preceq_{\mathrm{OWA}}$ and $\preceq_{\mathrm{CWA}}$, we can test relations $\preceq^+_{\mathrm{IT}}$ and $\preceq^-_{\mathrm{IT}}$ in polynomial (quadratic) time.

A set A of incomplete tuples can be viewed as a formula (which we also denote A, using the duality between tuples and formulae), which is the conjunction of all $\bar{a}$ in A. Likewise, we can also look at conjunction of all formulae $\neg\bar{a}$, giving us a formula $A^\neg$. That is, positive and negative formulae associated with A are:

$$A \ = \ \bigwedge \{\bar{a} \mid \bar{a} \in A\} \qquad A^\neg \ = \ \bigwedge \{\neg\bar{a} \mid \bar{a} \in A\} \tag{7}$$

Note that the first equation simply extends the duality of incomplete tuples and formulae to sets of incomplete tuples: it just tells us how to view a set A as a formula.

The following connection between orderings on sets and entailment of formulae is easily obtained from the definitions and containment criteria for conjunctive queries and their unions.

Proposition 2. $A \preceq_{\mathrm{IT}}^{+} B$ iff $B \supset A$, and $A \preceq_{\mathrm{IT}}^{-} B$ iff $A^{\neg} \supset B^{\neg}$.

Equipped with this, we can show that query answering by means of finding positive and negative incomplete tuples, i.e., by using (6), is always possible and preserves informativeness when input databases are interpreted under OWA or CWA.

Theorem 3. *Assume that input databases are interpreted under* OWA *or* CWA, *and that* $\mathbb{F}^{+}$ *consists of incomplete tuples, and* $\mathbb{F}^{-}$ *consists of their negations. Then for every query* Q *and every database* D *there exist finite sets* $Q_{\square}^{+}(D)$ *and* $Q_{\square}^{-}(D)$ *of incomplete tuples that, when viewed as formulae (7), are equivalent to* $\square^{+} Q(\llbracket D \rrbracket)$ *and* $\square^{-} Q(\llbracket D \rrbracket)$:

$$\mathsf{Mod}(Q_{\square}^{+}(D)) = \mathsf{Mod}(\mathsf{Th}^{+}(Q(\llbracket D \rrbracket)))$$
$$\mathsf{Mod}(Q_{\square}^{-}(D)^{\neg}) = \mathsf{Mod}(\mathsf{Th}^{-}(Q(\llbracket D \rrbracket)))$$

Moreover, this way of query answering preserves informativeness: if $D \preceq D'$ *(under the ordering given by the* CWA *or the* OWA *semantics), then*

$$Q_{\square}^{+}(D) \preceq_{\mathrm{IT}}^{+} Q_{\square}^{+}(D') \quad and \quad Q_{\square}^{-}(D) \preceq_{\mathrm{IT}}^{-} Q_{\square}^{-}(D').$$

6 Certain Information via Incomplete Tuples

The conclusion of the previous section is that incomplete tuples are a good representational mechanism for query answers over incomplete relational databases. What makes them especially suitable for the task is the *duality* of incomplete tuples: each one can be viewed both as a formula $\exists \bar{x} A(\bar{a}, \bar{x})$, satisfied by the query answer, or as an actual tuple $(\bar{x}, \bar{a})$, where $\bar{x}$ is a tuple of nulls. Thus, a set of tuples can be seen both as sets of formulae (7) representing our knowledge (positive and negative) about query answers, and an actual database relation with nulls. This duality lets us compute such knowledge using well established database query evaluation techniques, and present it to the user in a familiar format.

Ideally, following Theorem 3, we want to compute, for a query Q and a database D, sets $Q_{\square}^{+}(D)$ and $Q_{\square}^{-}(D)$ of incomplete tuples such that $Q_{\square}^{+}(D)$ is equivalent to $\square^{+} Q(\llbracket D \rrbracket)$ and $Q_{\square}^{-}(D)^{\neg}$ is equivalent to $\square^{-} Q(\llbracket D \rrbracket)$. That is,

$$\mathsf{Mod}(\bigwedge \{\bar{a} \mid \bar{a} \in Q_{\square}^{+}(D)\}) = \mathsf{Mod}(\mathsf{Th}^{+}(Q(\llbracket D \rrbracket))) \quad and$$
$$\mathsf{Mod}(\bigwedge \{\neg \bar{a} \mid \bar{a} \in Q_{\square}^{-}(D)\}) = \mathsf{Mod}(\mathsf{Th}^{-}(Q(\llbracket D \rrbracket)))$$

This is problematic even for first-order queries, however, as computing such sets of incomplete tuples is expensive. In fact, a simple examination of proofs in [1,18] shows that even when Q is a fixed Boolean FO query, checking whether $\square^{+} Q(\llbracket D \rrbracket_{\mathrm{CWA}})$ is true is CONP-complete, and the same question for $\square^{+} Q(\llbracket D \rrbracket_{\mathrm{OWA}})$ is undecidable.

But the discussion following the definition (6) showed a way out of this problem: we need to compute approximate answers with some guarantees, that is, formulae from positive and negative theories of $Q(\llbracket D \rrbracket)$. Using the duality of incomplete tuples, we say that, for a query Q, the pair (Q^+, Q^-) of queries returning sets of incomplete tuples provides a *sound answer* for Q under $\llbracket \ \rrbracket$ if, for every database D,

$$Q^+(D) \subseteq \mathsf{Th}^+(Q(\llbracket D \rrbracket)) \quad \text{and} \quad Q^-(D) \subseteq \mathsf{Th}_-^+(Q(\llbracket D \rrbracket)). \qquad (8)$$

Indeed, $Q^+(D)$ and $Q^-(D)$ represent parts of certain positive and negative knowledge about $Q(\llbracket D \rrbracket)$. If furthermore they can be computed with tractable data complexity, we say that they provide an *efficient sound answer* to Q on D.

Note that the right way to read sound answers is *tuple-by-tuple*: for instance, if $(\bot, 1)$ and $(\bot, 2)$ are in $Q^+(D)$, the correct interpretation is that for every $D' \in \llbracket D \rrbracket$, the answer $Q(D')$ contains a tuple whose second component is 1, and a tuple whose second component is 2. It is not meant to say that the first components of such tuples are the same: incomplete tuples cannot make cross-tuple statements.

Efficient Sound Answers Under OWA and CWA. There are trivial ways of finding sound answers: for instance, by letting Q^+ and Q^- return the empty set. Of course this is not what we want; instead we would like to find a good approximation of positive and negative certain information. To find the exact representation of such information, or a representation with some quality guarantees, and to do so efficiently, is impossible due to the complexity considerations explained earlier (which apply even to Boolean queries).

Thus, we shall present one particular inductive definition of queries Q^+ and Q^- that provides efficient sound answers for the most commonly used semantics of incompleteness, i.e., OWA and CWA semantics, for all FO queries. We also assume, as is standard in the database context, that they are evaluated under the active domain semantics, i.e., the answer to a k-ary query $Q(\bar{x})$ on D, denoted by $Q(D)$, is the set of tuples $\bar{a} \in \mathsf{adom}(D)^k$ so that $D \models Q(\bar{a})$. Formulae Q^+ and Q^- will use additional atomic formulae $\mathsf{const}(x)$ saying that x is not a null, i.e., an element of Const. We also write $\mathsf{const}(x_1, \ldots, x_n)$ for the conjunction of all $\mathsf{const}(x_i)$ for $1 \leq i \leq n$.

The definitions of Q^+ and Q^- are identical for OWA and CWA, except in the case of relational atomic formulae. We now present them inductively for the following formulae constructors: $Q(\bar{x}, \bar{y}, \bar{z}) = Q_1(\bar{x}, \bar{y}) \wedge Q_2(\bar{x}, \bar{z})$ (to account properly for the use of free variables in conjuncts); $Q(\bar{x}, \bar{y}, \bar{z}) = Q_1(\bar{x}, \bar{y}) \vee Q_2(\bar{x}, \bar{z})$ (likewise for disjunction); $Q(\bar{x}) = \neg Q_1(\bar{x})$; and $Q(\bar{x}) = \exists y Q_1(\bar{x}, y)$; as well as equational atoms $x = y$ and $x = a$ for a constant $a \in \mathsf{Const}$.

– If $Q(\bar{x}, \bar{y}, \bar{z}) = Q_1(\bar{x}, \bar{y}) \wedge Q_2(\bar{x}, \bar{z})$, then

$$Q^+(\bar{x}, \bar{y}, \bar{z}) = Q_1^+(\bar{x}, \bar{y}) \wedge Q_2^+(\bar{x}, \bar{z}) \wedge \mathsf{const}(\bar{x})$$
$$Q^-(\bar{x}, \bar{y}, \bar{z}) = Q_1^-(\bar{x}, \bar{y}) \vee Q_2^-(\bar{x}, \bar{z})$$

– If $Q(\bar{x}, \bar{y}, \bar{z}) = Q_1(\bar{x}, \bar{y}) \vee Q_2(\bar{x}, \bar{z})$, then

$$Q^+(\bar{x}, \bar{y}, \bar{z}) = Q_1^+(\bar{x}, \bar{y}) \vee Q_2^+(\bar{x}, \bar{z})$$
$$Q^-(\bar{x}, \bar{y}, \bar{z}) = Q_1^-(\bar{x}, \bar{y}) \wedge Q_2^-(\bar{x}, \bar{z})$$

– If $Q(\bar{x}) = \neg Q_1(\bar{x})$, then $Q^+(\bar{x}) = Q_1^-(\bar{x})$ and $Q^-(\bar{x}) = Q_1^+(\bar{x}) \wedge \mathsf{const}(\bar{x})$.
– If $Q(\bar{x}) = \exists y Q_1(\bar{x}, y)$, then $Q^+(\bar{x}) = \exists y Q_1^+(\bar{x}, y)$ and $Q^-(\bar{x}) = \forall y Q_1^-(\bar{x}, y)$.
– If $Q(x) = (x = a)$, then $Q^+(x) = (x = a)$ and $Q^-(x) = \neg(x = a) \wedge \mathsf{const}(x)$.
– If $Q(x, y) = (x = y)$, then

$$Q^+(x, y) = (x = y) \quad \text{and} \quad Q^-(x, y) = \neg(x = y) \wedge \mathsf{const}(x, y).$$

Note that the rules for $\wedge$ and $\vee$ are not symmetric, due to the asymmetric rule for negation.

Finally we define such queries for atomic formulae $R(\bar{x})$, when R is a database relation, as follows:

Under OWA: $R^+(\bar{x}) = R(\bar{x})$ $R^-(x, y) = \mathsf{false}$

Under CWA: $R^+(\bar{x}) = R(\bar{x})$ $R^-(x, y) = \neg\exists \bar{y}(R(\bar{y}) \wedge \alpha_\lhd(\bar{x}, \bar{y}))$

Here we use an additional formula $\alpha_\lhd(\bar{x}, \bar{y})$ such that $\alpha_\lhd(\bar{a}, \bar{b})$ iff $\bar{a} \lhd \bar{b}$. It is not hard to see that it can be defined as a quantifier-free formula that uses equalities and $\mathsf{const}(\cdot)$, as a disjunction over possible instantiations of variables $\bar{x}, \bar{y}$ as constants or nulls. These give us complete definitions of Q^+ and Q^- under OWA and CWA.

Theorem 4. *The definitions of Q^+ and Q^- above provide efficient sound answers to FO queries under OWA and CWA. The data complexity of such queries is in AC^0.*

Example. Consider the difference query $Q(\bar{x}) = R(\bar{x}) \wedge \neg S(\bar{x})$ that is among the most troublesome operations for relational query evaluation with nulls [14,22,29].

Then the query $Q^+(\bar{x})$ is $R(\bar{x}) \wedge S^-(\bar{x}) \wedge \mathsf{const}(\bar{x})$. Thus, under OWA, S^- and hence Q^+ is equivalent to false, which is to be expected, as under OWA the difference query returns the empty set. Under CWA, on the other hand, Q^+ computes the set of constant tuples in R which do not match any tuple in S.

With Q^-, we can also infer useful negative knowledge. Applying the rules, $Q^-(\bar{x}) = R^-(\bar{x}) \vee (S(\bar{x}) \wedge \mathsf{const}(\bar{x}))$. Thus, under OWA it becomes $S(\bar{x}) \wedge \mathsf{const}(\bar{x})$ and we get information that constant tuples in S will never be in the answer, something that traditional certain answers will miss. Under CWA, we also see that tuples not mapped into tuples of R (i.e., R^-) can never be query answers.

These are exactly the query results one would expect, and they are obtained by a direct application of transformations giving us queries Q^+ and Q^-.

7 Conclusion

When answering queries over incomplete data, one should concentrate not only on what is guaranteed to be true, but also on what is guaranteed to be false, i.e., negative information. Finding such negative information however is often ignored. We showed how to apply the framework for dealing with incompleteness based on semantics, knowledge, and ordering, to define negative information that can with certainty be inferred about query answers. We showed how to use basic properties of such negative information to find a good representational mechanism for relational query answering, resulting in a natural, but hitherto not widely used mechanism of incomplete tuples. To prove its applicability, we demonstrated an efficient procedure for computing positive and negative knowledge for all FO queries over relational databases.

As next steps, we would like to see how these notions behave in standard applications of incompleteness (integration, inconsistency, etc.), relate them to other approximate query answering notions, both in databases [15,16,23,34] and in AI [27,32], and to existing approaches that explain why tuples do not appear in query answers [21,36]. As for quality of approximations of certain answers, these are best confirmed experimentally, as was demonstrated recently [19].

Acknowledgments. I am grateful to anonymous referees for their comments. This work was partly supported by EPSRC grants J015377 and M025268.

References

1. Abiteboul, S., Kanellakis, P., Grahne, G.: On the representation and querying of sets of possible worlds. Theoret. Comput. Sci. **78**(1), 158–187 (1991)
2. Abiteboul, S., Hull, R., Vianu, V.: Foundations of Databases. Addison-Wesley, Reading (1995)
3. Abiteboul, S., Segoufin, L., Vianu, V.: Representing and querying XML with incomplete information. ACM TODS **31**(1), 208–254 (2006)
4. Arenas, M., Barceló, P., Libkin, L., Murlak, F.: Foundations of Data Exchange. Cambridge University Press, Cambridge (2014)
5. Barceló, P., Libkin, L., Poggi, A., Sirangelo, C.: XML with incomplete information. J. ACM **58**(1), 238–272 (2010)
6. Barceló, P., Libkin, L., Reutter, J.: Querying regular graph patterns. J. ACM **61**(1), 1–54 (2014)
7. Bertossi, L.: Database Repairing and Consistent Query Answering. Morgan & Claypool Publishers, San Rafael (2011)
8. Bienvenu, M., ten Cate, B., Lutz, C., Wolter, F.: Ontology-based data access: a study through disjunctive datalog, CSP, and MMSNP. ACM TODS **39**(4), 1–44 (2014)
9. Buneman, P., Jung, A., Ohori, A.: Using power domains to generalize relational databases. Theoret. Comput. Sci. **91**(1), 23–55 (1991)
10. Calì, A., Lembo, D., Rosati, R.: On the decidability and complexity of query answering over inconsistent and incomplete databases. In: PODS, pp. 260–271 (2003)

11. Calvanese, D., De Giacomo, G., Lembo, D., Lenzerini, M., Rosati, R.: Tractable reasoning and efficient query answering in description logics: the DL-Lite family. J. Autom. Reasoning **39**(3), 385–429 (2007)

12. Compton, K.: Some useful preservation theorems. J. Symbol. Logic **48**(2), 427–440 (1983)

13. Console, M., Guagliardo, P., Libkin, L.: Approximations and refinements of certain answers via many-valued logics. In: KR 2016, pp. 349–358 (2016)

14. Date, C.J., Darwen, H.: A Guide to the SQL Standard. Addison-Wesley, Reading (1996)

15. Fink, R., Olteanu, D.: On the optimal approximation of queries using tractable propositional languages. In: ICDT, pp. 174–185 (2011)

16. Garofalakis, M., Gibbons, P.: Approximate query processing: taming the terabytes. In: VLDB (2001)

17. Gheerbrant, A., Libkin, L., Sirangelo, C.: Naïve evaluation of queries over incomplete databases. ACM TODS **39**(4), 231 (2014)

18. Gheerbrant, A., Libkin, L.: Certain answers over incomplete XML documents: extending tractability boundary. Theory Comput. Syst. **57**(4), 892–926 (2015)

19. Guagliardo, P., Libkin, L.: Making SQL queries correct on incomplete databases: a feasibility study. In: PODS 2016, pp. 211–223 (2016)

20. Gunter, C.: Semantics of Programming Languages. The MIT Press, Cambridge (1992)

21. Herschel, M., Hernández, M.: Explaining missing answers to SPJUA queries. PVLDB **3**(1), 185–196 (2010)

22. Imielinski, T., Lipski, W.: Incomplete information in relational databases. J. ACM **31**(4), 761–791 (1984)

23. Ioannidis, Y.: Approximations in database systems. In: Calvanese, D., Lenzerini, M., Motwani, R. (eds.) ICDT 2003. LNCS, vol. 2572, pp. 16–30. Springer, Heidelberg (2002)

24. Klein, H.: On the use of marked nulls for the evaluation of queries against incomplete relational databases. In: Polle, T., Ripke, T., Schewe, K. (eds.) Fundamentals of Information Systems, pp. 81–98. Springer, New York (1999)

25. Kontchakov, R., Lutz, C., Toman, D., Wolter, F., Zakharyaschev, M.: The combined approach to ontology-based data access. In: IJCAI, pp. 2656–2661 (2011)

26. Lenzerini, M.: Data integration: a theoretical perspective. In: ACM Symposium on Principles of Database Systems (PODS), pp. 233–246 (2002)

27. Levesque, H.: A completeness result for reasoning with incomplete first-order knowledge bases. In: KR, pp. 14–23 (1998)

28. Libkin, L.: Certain answers as objects and knowledge. Artif. Intell. **232**, 1–19 (2016)

29. Libkin, L.: SQL's three-valued logic and certain answers. ACM TODS **41**(1), 1–28 (2016)

30. Lipski, W.: On semantic issues connected with incomplete information databases. ACM TODS **4**(3), 262–296 (1979)

31. Lipski, W.: On relational algebra with marked nulls. In: PODS 1984, pp. 201–203 (1984)

32. Liu, Y., Levesque, H.: A tractability result for reasoning with incomplete first-order knowledge bases. In: IJCAI, pp. 83–88 (2003)

33. Reiter, R.: Towards a logical reconstruction of relational database theory. In: Brodie, M.L., Mylopoulos, J., Schmidt, J.W. (eds.) On Conceptual Modelling, pp. 191–233. Springer, New York (1982)

34. Reiter, R.: A sound and sometimes complete query evaluation algorithm for relational databases with null values. J. ACM **33**(2), 349–370 (1986)
35. Rosati, R.: On the decidability and finite controllability of query processing in databases with incomplete information. In: PODS, pp. 356–365 (2006)
36. Shmueli, O., Tsur, S.: Logical diagnosis of LDL programs. In: ICLP, pp. 112–129 (1990)

Extending Weakly-Sticky Datalog$^{\pm}$: Query-Answering Tractability and Optimizations

Mostafa Milani$^{(\boxtimes)}$ and Leopoldo Bertossi

School of Computer Science, Carleton University, Ottawa, Canada
{mmilani,bertossi}@scs.carleton.ca

Abstract. *Weakly-sticky* (*WS*) Datalog$^{\pm}$ is an expressive member of the family of Datalog$^{\pm}$ programs that is based on the syntactic notions of *stickiness* and *weak-acyclicity*. Query answering over the *WS* programs has been investigated, but there is still much work to do on the design and implementation of practical query answering (QA) algorithms and their optimizations. Here, we study sticky and *WS* programs from the point of view of the behavior of the chase procedure, extending the stickiness property of the chase to that of *generalized stickiness of the chase (gsch-property)*. With this property we specify the semantic class of *GSCh* programs, which includes sticky and *WS* programs, and other syntactic subclasses that we identify. In particular, we introduce *joint-weakly-sticky* (*JWS*) programs, that include *WS* programs. We also propose a bottom-up QA algorithm for a range of subclasses of *GSCh*. The algorithm runs in polynomial time (in data) for *JWS* programs. Unlike the *WS* class, *JWS* is closed under a general magic-sets rewriting procedure for the optimization of programs with existential rules. We apply the magic-sets rewriting in combination with the proposed QA algorithm for the optimization of QA over *JWS* programs.

1 Introduction

Ontology-based data access (OBDA) [23] allows to access, through a conceptual layer that takes the form of an ontology, underlying data that is usually stored in a relational database. Queries can be expressed in terms of the ontology language, but are answered by eventually appealing to the extensional data underneath. Common languages of choice for representing ontologies are certain classes (or fragments) of *description logic* (DL) [3] and, more recently, of *Datalog*$^{\pm}$ [8,10]. Those classes are expected to be computationally well-behaved in relation to query answering (QA). Several approaches for QA, and a number of techniques have been proposed for DL-based [3,23] and Datalog$^{\pm}$-based OBDA [8]. In this work we concentrate on the conjunctive QA problem from relational data through Datalog$^{\pm}$ ontologies.

Datalog$^{\pm}$, as an extension of the Datalog query language [11], allows in rule heads (i.e. consequents): existentially quantified variables ($\exists$-variables), equality atoms, and a false propositional atom, say **false**, to represent "negative program

© Springer International Publishing Switzerland 2016
M. Ortiz and S. Schlobach (Eds.): RR 2016, LNCS 9898, pp. 128–143, 2016.
DOI: 10.1007/978-3-319-45276-0_10

constraints" [8–10]. Hence the "+" in Datalog$^\pm$, while the "−" reflects syntactic restrictions on programs for better computational properties.

Datalog$^\pm$ is expressive enough to represent in logical and declarative terms useful ontologies, in particular those that capture and extend the common conceptual data models [9] and Semantic Web data [2]. The rules of a Datalog$^\pm$ program can be seen as forming an ontology on top of an extensional database, D, which may be *incomplete*. In particular, the ontology: (a) provides a "query layer" for D, enabling OBDA, and (b) specifies a completion of D.

In the rest of this work we will assume that programs contain only existential rules (plus extensional data). When programs are subject to syntactic restrictions, we talk about Datalog$^\pm$ programs, whereas when no conditions are assumed or applied, we talk about Datalog$^+$ programs, also called Datalog$^\exists$ programs [4,8,15,16].

From the semantic and computational point of view, the completion of the underlying extensional instance D appeals to so-called *chase* procedure that, starting from D, iteratively enforces the rules in the ontology. That is, when a rule body (the antecedent) becomes true in the instance so far, but not the head (the consequent), a new tuple is generated. This process may create new values (nulls) or propagate values to the same or other *positions*. The latter correspond to the arguments in the schema predicates.

Example 1. Consider a Datalog$^\pm$ program $\mathcal{P}$ with extensional database $D = \{r(a, b)\}$ and set of rules $\mathcal{P}^r$:

$$r(X, Y) \rightarrow \exists Z \; r(Y, Z). \tag{1}$$

$$r(X, Y), r(Y, Z) \rightarrow s(X, Y, Z). \tag{2}$$

The positions for this schema are: $r[1], r[2], s[1], s[2], s[3]$. The extension of D generated by the chase includes the following tuples (among infinitely many others): $r(b, \zeta_1), s(a, b, \zeta_1), r(\zeta_1, \zeta_2), s(b, \zeta_2, \zeta_1)$. Notice that $s(a, b, \zeta_1)$ and $s(b, \zeta_1, \zeta_2)$ are obtained by replacing the *join variable* Y (i.e. repeated) in the body of (2) by b and ζ_1, resp. ∎

The result of the chase, seen as an instance for the combined ontological and relational schema, is also called "the chase". The chase (instance) extends D, but may be infinite; and gives the semantics to the Datalog$^\pm$ ontology, by providing an intended model, and can be used for QA. At least conceptually, the query can be posed directly to the materialized chase instance. However, this may not be the best way to go about QA, and computationally better alternatives have to be explored.

Actually, when the chase is infinite, (conjunctive) QA may be undecidable [14]. However, in some cases, even with an infinite chase, QA is still computable (decidable), and even tractable in the size of D. In fact, syntactically restricted subclasses of Datalog$^+$ programs have been identified and characterized for which QA is decidable, among them: *linear, guarded* and *weakly-guarded, sticky* and *weakly-sticky* (*WS*) [8,10] Datalog$^\pm$.

Sticky Datalog$^\pm$ is a syntactic class of programs characterized by syntactic restrictions on join variables. *WS* Datalog$^\pm$ extends sticky Datalog$^\pm$ by also capturing the well-known class of *weakly-acyclic programs* [13], which is defined in terms of the syntactic notions of *finite-* and *infinite-rank* positions. Accordingly, *WS* Datalog$^\pm$ is characterized by restrictions on join variables occurring in infinite-rank positions. A non-deterministic QA algorithm for *WS* Datalog$^\pm$ is presented in [10], to establish the theoretical result that QA can be done in polynomial-time in data.

In this work, we concentrate on sticky and *WS* Datalog$^\pm$, because they have found natural applications in our previous work on extraction of quality data from possible dirty databases [20]. The latter task is accomplished through QA, so that the need for efficient QA algorithms becomes crucial. Accordingly, the main motivations, goals, and results (among others) for/in this work are:

(A) Providing a practical, bottom-up QA algorithm for *WS* Datalog$^\pm$. Being bottom-up, it is expected to be based on (a variant of) the chase. Since the latter can be infinite, the query at hand guarantees that the need to generate only an initial, finite portion of the chase.

(B) Optimizing the QA algorithm through a *magic-sets* rewriting technique, to make it more query sensitive.

For (B), we apply the magic-sets technique for Datalog$^+$ first introduced in [1], which we denote with MagicD$^+$. Extending classical magic-sets for Datalog [11], MagicD$^+$ prevents existential variables from getting bounded, a reasonable adjustment that essentially preserves the semantics of existential rules during the rewriting. Unfortunately, the class of *WS* Datalog$^\pm$ programs is provably not closed under MagicD$^+$, meaning that the result of applying MagicD$^+$ to a *WS* program may not be *WS* anymore. This led us to search for a more general class of programs that is: (i) closed under MagicD$^+$, (ii) extends *WS* Datalog$^\pm$, and (iii) has an efficient QA algorithm. Notice that at this point both syntactic and semantic classes may be investigated, and we do so. The latter classes refer to the properties of the chase as an instance.

Sticky programs enjoy the *stickiness property of the chase*, which -in informal terms- means the following: If, due to the application of a rule during the chase, a value replaces a join variable in the rule body, then that value is propagated through all the possible subsequent steps, i.e. the value "sticks". The "stickiness property of the chase" defines a "semantic class", *SCh*, in the sense that it is characterized in terms of the chase for programs that include an extensional database. This class properly extends sticky Datalog$^\pm$ [10].

We can relax the condition in the *sch-property*, and define the *generalized-stickiness property of the chase*. It is as for the *sch-property*, but with the propagation condition only on join variables that *do not* appear in the *finite positions*; the latter being those where finitely many different values may appear during the chase. With this property we define the new semantic class of *GSCh* programs. However, we make notice that, given a program $\mathcal{P}$ consisting of a set of rules $\mathcal{P}^r$ and an extensional instance D, computing (deciding) *FinPoss*($\mathcal{P}$), the set of finite positions of $\mathcal{P}$, is unsolvable (undecidable) [12]. Accordingly, it is also undecidable if a Datalog$^+$ program belongs to the *GSCh* class.

Starting from the definition of the *GSCh* class, we can define, backwardly, a whole range of different semantic classes between *Sticky* and *GSCh*, by replacing in the definition of the latter the condition on the set of non-finite positions by a stronger one that appeals to a superset of them. Each of these supersets is represented through its complement, which is determined by an abstract *selection function* $\mathcal{S}$ that identifies a set of finite positions. Such a function, given a program $\mathcal{P}$, returns a subset $\mathcal{S}(\mathcal{P})$ of $FinPoss(\mathcal{P})$ (making $\mathcal{S}$ sound, but possibly incomplete w.r.t $FinPoss(\mathcal{P})$). $\mathcal{S}$ may be computable or not, and may depend on $\mathcal{P}^r$ alone or on the combination of $\mathcal{P}^r$ and D. Hence we split $\mathcal{P}$ into $\mathcal{P}^r$ and D. The corresponding semantic class of programs, those enjoying the $\mathcal{S}$-*stickiness property of the chase*, is denoted with $SCh(\mathcal{S})$.

In particular, if $\mathcal{S}^\top$ is the non-computable function that selects all finite positions, $GSCh = SCh(\mathcal{S}^\top)$. If $\mathcal{S}^{rank}$ selects the finite-rank positions (that happen to be finite positions) [13], then $WSCh = SCh(\mathcal{S}^{rank})$ is a new semantic class programs, those with the *weak-stickiness property of the chase*. And for the class *SCh* of programs we started from above, it holds $SCh = SCh(\mathcal{S}^\perp)$, with $\mathcal{S}^\perp$ always returning the empty set of positions. Notice that $\mathcal{S}^{rank}$ and $\mathcal{S}^\perp$ are both computable, and they do not use the extensional instance D, but only the program. In this sense, we say that they are *syntactic selection functions*.

We can see that the combination of selection functions with the $\mathcal{S}$-based notion of stickiness property of the chase (i.e. that only values in join variables in positions outside those selected by $\mathcal{S}$ propagate all the way through), defines a range of semantic classes of programs starting with *SCh*, ending with *GSCh*, and with $SCh(\mathcal{S}^{rank})$ in between. They are shown in ascending order of inclusion, from left to right, in the middle layer of Fig. 1. There, the upper layer shows the corresponding selection functions ordered by inclusion (of their images).

$$\mathcal{S}^\perp \quad \underset{(a)}{\subsetneq} \quad \mathcal{S}^{rank} \quad \underset{(b)}{\subsetneq} \quad \mathcal{S}^\exists \quad \underset{(c)}{\subsetneq} \quad \mathcal{S}^\top$$

$$\uparrow \qquad\qquad \uparrow \qquad \uparrow \qquad\qquad \uparrow$$

$$SCh{:=}SCh(\mathcal{S}^\perp) \underset{(d)}{\subsetneq} WSCh{:=}SCh(\mathcal{S}^{rank}) \underset{(e)}{\subsetneq} SCh(\mathcal{S}^\exists) \underset{(f)}{\subsetneq} GSCh{:=}SCh(\mathcal{S}^\top)$$

$$\underset{(g)}{\cup\!\!\!\!\!\!\!\ast} \qquad\qquad \underset{(h)}{\cup\!\!\!\!\!\!\!\ast} \qquad\qquad \underset{(i)}{\cup\!\!\!\!\!\!\!\ast}$$

$$Sticky \quad \underset{(j)}{\subsetneq} \quad WS \quad \underset{(k)}{\subsetneq} \quad JWS$$

Fig. 1. Semantic and syntactic program classes, and selection functions

A parallel and corresponding range of syntactic classes, also ordered by set inclusion, is shown in the lower layer. It includes the sticky and *WS* classes (cf. Fig. 1, bottom). Each syntactic class only partially represents its semantic counterpart, in the sense that the former: does not consider extensional instances, appeals to the same selection function, but also imposes additional syntactic conditions on the set of rules. All the inclusions in Fig. 1 are proper, as examples we provide in this work will show (but (g) and (j) are known [10]).

In this work, our main goal is to introduce and investigate the semantic class $SCh(\mathcal{S}^\exists)$, determined by the selection function $\mathcal{S}^\exists$ that is defined in terms of the *existential dependency graph* of a program [15] (a syntactic, computable construction). We also introduce and investigate its corresponding syntactic class of *joint-weakly-sticky (JWS)* programs. The latter happens to satisfy desiderata (A) and (B) above. Actually, about (A), we provide for the class $SCh(\mathcal{S}^\exists)$ a polynomial-time, chase-based, bottom-up QA algorithm, which can be applied to *JWS* (and all its semantic and syntactic subclasses) in particular. This is a general situation: The polynomial-time QA algorithms for the classes *Sticky* [10], *WS* [10,21], and *JWS* (this work) rely basically on the properties of the semantic class rather than on the specific syntactic restrictions. Hence our interest is in investigating the particular semantics classes, and semantic classes in general, as defined by selection functions. About (B), notice that if we start with a *WS* program, we can apply MagicD$^+$ to it, obtaining a *JWS* program, for which QA can be done in polynomial time.

The paper is structured as follows: Sect. 2 is a review of some basics of the database theory, the chase procedure, and Datalog$^\pm$. Section 3 contains the definition of the stickiness and general-stickiness properties of the chase and the *SCh* and *GSCh* semantic classes. Section 4 is about the ranges of syntactic and semantic program subclasses of *GSCh*. The *JWS* class of programs is introduced in Sect. 5. Sections 6 and 7 contain the QA algorithm and MagicD$^+$. In this paper we use mainly intuitive and informal introductions of concepts and techniques, illustrated by examples. The precise technical developments can be found in the Appendices of [22].

2   Preliminaries

We start with a relational schema $\mathcal{R}$ containing two disjoint "data" sets: $\mathcal{C}$, a possibly infinite domain of *constants*, and $\mathcal{N}$, of infinitely many *labeled nulls*. It also contains predicates of fixed and finite arities. If p is an n-ary predicate (i.e. with n arguments) and $1 \leq i \leq n$, $p[i]$ denotes its i-th position. With $\mathcal{R}$, $\mathcal{C}$, $\mathcal{N}$ we can build a language $\mathcal{L}$ of first-order (FO) predicate logic, that has $\mathcal{V}$ as its infinite set of *variables*. We denote with $\bar{X}$, etc., finite sequences of variables. A *term* of the language is a constant, a null, or a variable. An *atom* is of the form $p(t_1, \ldots, t_n)$, with $p \in \mathcal{R}$, n-ary predicate, and $t_1, \ldots, t_n$ terms. An atom is *ground*, if it contains no variables. An *instance* I for schema $\mathcal{R}$ is a possibly infinite set of ground atoms. The *active domain* of an instance I, denoted $Adom(I)$, is the set of constants or nulls that appear in I. Instances can be used as interpretation structures for the FO language $\mathcal{L}$. Accordingly, we can use the notion of formula satisfaction of FO predicate logic.

A conjunctive query (CQ) is a FO formula, $\mathcal{Q}(\bar{X})$, of the form: $\exists \bar{Y}(p_1(\bar{X}_1) \wedge \cdots \wedge p_n(\bar{X}_n))$, with $\bar{Y} := (\bigcup \bar{X}_i) \setminus \bar{X}$. For an instance I, $\bar{t} \in (\mathcal{C} \cup \mathcal{N})^n$ is an *answer* to $\mathcal{Q}$ if $I \models \mathcal{Q}[\bar{t}]$, with $\bar{t}$ replacing the variables in $\bar{X}$. $\mathcal{Q}(I)$ denotes the set of answers to $\mathcal{Q}$ in I. $\mathcal{Q}$ is Boolean (a BCQ) when $\bar{X}$ is empty, and when true in I, $\mathcal{Q}(I) := \{yes\}$. Otherwise, $\mathcal{Q}(I) = \emptyset$. Notice that a CQ can be expressed as a rule of the form $p_1(\bar{X}_1), ..., p_n(\bar{X}_n) \rightarrow ans_{\mathcal{Q}}(\bar{X})$, where $ans_{\mathcal{Q}}(\cdot) \notin \mathcal{R}$ is an auxiliary predicate. The query answers form the extension of the answer-collecting predicate $ans_{\mathcal{Q}}(\cdot)$.[1]

A *tuple-generating dependency* (*TGD*), also called *existential rule* or simply a *rule* is a sentence, σ, of $\mathcal{L}$ of the form: $p_1(\bar{X}_1), \ldots, p_n(\bar{X}_n) \rightarrow \exists \bar{Y} p(\bar{X}, \bar{Y})$, with $\bar{X}_i$ indicating the variables appearing in p_i (among possibly elements from $\mathcal{C}$), and an implicit universal quantification over all variables in $\bar{X}_1, \ldots, \bar{X}_n, \bar{X}$, and $\bar{X} \subseteq \bigcup_i \bar{X}_i$, and the dots in the antecedent standing for conjunctions.[2] The variables in $\bar{Y}$, that could be empty, are *existential variables*. With $head(\sigma)$ and $body(\sigma)$ we denote the sets of atoms in the consequent and the antecedent of σ, respectively. The notions of satisfaction by an instance I of a TGD σ (denoted $I \models \sigma$), and of a set of TGDs, are defined as in FO logic.

A Datalog$^+$ program $\mathcal{P}$ consists of a set of rules $\mathcal{P}^r$ and an extensional database instance D, i.e. a finite instance whose atoms contain only elements from $\mathcal{C}$. The set of models of $\mathcal{P}$, denoted by $Mod(\mathcal{P})$, contains all instances I, such that $I \supseteq D$ and $I \models \mathcal{P}^r$. Given a CQ $\mathcal{Q}$, the set of answers to $\mathcal{Q}$ from $\mathcal{P}$ is defined by $ans(\mathcal{Q}, \mathcal{P}) := \bigcap_{I \in Mod(\mathcal{P})} \mathcal{Q}(I)$.

The *chase* procedure is a fundamental algorithm in different database problems, including implication of database dependencies, query containment, and CQ answering under dependencies [6,10,13,14,17]. For the latter problem [10,13], the idea is that, given a set of dependencies over a database schema and an instance as input, the chase enforces the dependencies by adding new tuples into the instance, so that the result satisfies the constraints (cf. Appendix B in [22] for more details).

Example 2 (Example 1 cont.). With the given instance D and the assignment θ: $X \mapsto a, Y \mapsto b$, rule (1) is not satisfied: $D \models r(X,Y)[\theta]$, but $D \not\models \exists Z\, r(Y,Z)[\theta]$. Then, the chase inserts a new tuple $r(b, \zeta_1)$ into D (ζ_1 is a fresh null), resulting in instance D_1. D_1 does not satisfy (2), so the chase inserts $s(a, b, \zeta_1)$, resulting in instance D_2. The chase continues, without stopping, creating an infinite instance: $chase(\mathcal{P}) = \{r(a,b), r(b, \zeta_1), s(a, b, \zeta_1), r(b, \zeta_1), r(\zeta_1, \zeta_2), s(b, \zeta_1, \zeta_2), \ldots\}$. ∎

The instance resulting from the chase procedure is also called "the chase". As such, it is a so-called *universal model* [13], i.e. a representative of all models in $Mod(\mathcal{P})$. In particular, the answers to a CQ $\mathcal{Q}$ under $\mathcal{P}$, i.e. those in $ans(\mathcal{Q}, \mathcal{P})$, can be computed by evaluating $\mathcal{Q}$ over the chase (and discarding the answers containing nulls). The chase procedure may not terminate, and it is in general undecidable if it terminates, even for a fixed instance [12].

---

[1] When $\mathcal{Q}$ is Boolean, $ans_{\mathcal{Q}}$ is a propositional atom; and if $\mathcal{Q}$ is true in I, then $ans_{\mathcal{Q}}$ can be reinterpreted as the query answer.

[2] A query of this form can be seen and treated as a new TGD containing a fresh head predicate.

Several sufficient conditions, syntactic [12,13,18] and data-dependent [19], that guarantee chase termination have been identified. *Weak-acyclicity* [13] is one of the former, and is defined using the dependency graph.

Example 3 (Example 2 cont.). The *dependency graph* (*DG*) of $\mathcal{P}^r$ (cf. Fig. 2) is a directed graph whose vertices are the positions of $\mathcal{R}$.

The edges are defined as follows: for every $\sigma \in \mathcal{P}^r$, $\forall$-variable X in $head(\sigma)$, and position π in $body(\sigma)$: 1. for each occurrence of X in position π' in $head(\sigma)$, create an edge from π to π'. 2. for each $\exists$-variable Z in position π'' in $head(\sigma)$, create a *special edge* (dashed) from π to π''.

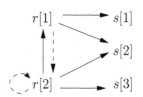

Fig. 2. Dependency graph

The *rank of a position* is the maximum number of special edges over all (finite or infinite) paths ending at that position. $\Pi_F(\mathcal{P}^r)$ is the set of finite-rank positions in $\mathcal{P}^r$. A program is *weakly-acyclic* (*WA*) if all of the positions have finite-rank. Here, $r[1], r[2] \notin \Pi_F(\mathcal{P}^r)$, so the program is not *WA*. ∎

In a program with finite- and infinite-rank positions, every finite-rank position is finite: For any extensional instance D, during the chase only polynomially many different values appear in them (in data) [10]. However, in infinite-rank positions, there may be infinitely many values (and the chase does not terminate). In particular, for every *WA* program and instance D the chase terminates in polynomially many steps with respect to the size of D [13].

The notions of finite and infinite positions mentioned above rely on the chase instance and hence a program's data: Given a program $\mathcal{P}$ with schema $\mathcal{R}$, the set of finite positions of $\mathcal{P}$, that we refer to as $FinPoss(\mathcal{P})$, is the set of positions where finitely many values appear in $chase(\mathcal{P})$. Every position that is not finite is infinite.

Conjunctive query answering w.r.t an arbitrary set of TGDs is in general undecidable [5]. The Datalog$^\pm$ family is formed by syntactic subclasses of Datalog$^+$ programs that are defined by imposing restrictions on the sets of TGDs rules in a program, to guarantee decidability, and in several cases, tractability of QA. In this work we concentrate on the sticky and *WS* classes of programs.

3   Stickiness of the Chase and its Generalization

The *"stickiness property of the chase"* (*sch-property*) [10] is a "semantic" property of Datalog$^+$ programs in relation to the way the chase behaves with the extensional data. We informally introduce it here. A program has this property if, due to the application of a rule σ, when a value replaces a repeated variable in a rule-body, then that value also appears in all the head atoms obtained through the iterative enforcement of applicable rules that starts with σ's application. In short, the value is propagated through all possible subsequent chase steps.

Example 4. Consider $\mathcal{P}_1$ with $D_1 = \{r(a,b), r(b,c)\}$, and $\mathcal{P}_1^r$ containing:

$$r(X,Y), r(Y,Z) \to p(Y,Z). \quad p(X,Y) \to \exists Z\, s(X,Y,Z). \quad s(X,Y,Z) \to u(Y).$$

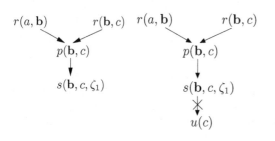

$\mathcal{P}_1$ does not have the *sch-property*, as the chase in Fig. 3 (right-hand side) shows: value b is not propagated all the way down to $u(c)$. However, a program $\mathcal{P}_2$ with the same database $D_2 = D_1$ but a set $\mathcal{P}_2^r$ of rules which is $\mathcal{P}_1^r$ without its third rule, has the *sch-property*, as shown in Fig. 3 (left-hand side). ∎

Fig. 3. The *sch-property*.

SCh is the semantic class of programs with the *sch-property*. Next, we briefly recall the classes of programs whose definitions are related to the *sch-property* and the SCh programs.

Sticky Programs. Sticky Datalog$^\pm$ is a syntactic class of programs that enjoy the *sch-property*, for any extensional database [10]. Its programs are characterized through a body variable *marking procedure* whose input is the set $\mathcal{P}^r$ of program rules (the data do not participate).

The procedure has two steps: (a) *Preliminary step*, for each $\sigma \in \mathcal{P}^r$ and variable $X \in body(\sigma)$, if there is an atom $A \in head(\sigma)$ where X does not appear, mark each occurrence of X in $body(\sigma)$, and (b) *Propagation step*, for each $\sigma \in \mathcal{P}^r$, if a marked variable in $body(\sigma)$ appears at position π, then for every $\sigma' \in \mathcal{P}^r$ (including σ), mark each occurrence of the variables in $body(\sigma')$ that appear in $head(\sigma')$ in the same position π.

$\mathcal{P}^r$ is *sticky* when, after applying the marking procedure, there is no rule with a marked variable appearing more than once in its body (notice that a variable never appears both marked and unmarked in a same body).

Example 5. The initial set of three rules, $\mathcal{P}^r$, is shown on the left-hand side below. The second rule already shows marked variables (with hat) after the preliminary step. The set of rules on the right-hand side are the result of whole marking procedure.

$$r(X,Y), p(X,Z) \to s(X,Y,Z). \qquad r(\hat{X},Y), p(\hat{X},\hat{Z}) \to s(X,Y,Z).$$
$$s(\hat{X},Y,\hat{Z}) \to u(Y). \qquad s(\hat{X},Y,\hat{Z}) \to u(Y).$$
$$u(X) \to \exists\, Y\, r(Y,X). \qquad u(X) \to \exists\, Y\, r(Y,X).$$

Variables X and Z in the first rule-body end up marked after the propagation step: they appear in the same rule's head, in marked positions ($s[1]$ and $s[3]$ in the body of the second rule). Accordingly, the set of rules is *not* sticky: X in the first rule's body is marked and occurs twice (in $r[1]$ and $p[1]$). ∎

With sticky programs, QA can be done in polynomial-time in data complexity [10]. A program with the *sch-property* may not be syntactically sticky. Actually, the *SCh* class can be extended to several larger, semantic, classes of programs that enjoy a form of the *sch-property* with the propagation condition during the chase only on values in certain forms of "infinite" positions. (We propose a new, syntactic class along these lines in Sect. 4). Something similar can be done with the class of sticky programs.

Weakly-Sticky (*WS*) Programs. This is a syntactic class that extends those of *WA* and sticky programs. Its characterization uses the above notions of finite-rank and marked variable: A set of rules $\mathcal{P}^r$ is *WS* if, for every rule in it and every repeated variable in its body, the variable is either non-marked or appears in some position in $\Pi_F(\mathcal{P}^r)$.

Example 6 (Example 5 cont.). $\mathcal{P}^r$ is *WS*, because $p[1] \in \Pi_F(\mathcal{P}^r)$; and X, the only repeated variable in a body (of the first rule), is marked, but in $p[1]$. ■

The *WS* condition guarantees tractability of QA, because CQs can be answered on an initial fragment of the chase whose size is polynomial in that of the extensional database. This relies on these facts: (a) Finite-rank positions can be saturated by polynomially many values in the size of the extensional database. (b) Stickiness for infinite-rank positions ensures that polynomially many values are required in them for answering a query at hand. In fact, stickiness for infinite positions makes the number of values required in them for QA polynomially depend on the number of values in finite-rank positions. So, both in finite and infinite-rank positions, polynomially many values are needed.

The above argument about QA is more general than as applied to *WS* programs. It can be applied with more general, syntactic and semantic, classes of programs that are characterized through the use of the stickiness condition on positions where infinitely many values may appear during the chase. *WS* programs are a special case, where those positions are with infinite-rank; and the stickiness is enforced by the syntactic variable-marking mechanism. Actually, we can make the general claim that the combination of finitely many values in finite positions plus chase-stickiness on infinite positions makes QA decidable.

Generalized Stickiness. The *generalized-stickiness of the chase (gsch-property)* is defined by relaxing the condition in the *sch-property*: the condition applies to values for the repeated body variables that do not appear in *finite positions*. *GSCh* is the semantic class of programs with the *gsch-property* (cf. Fig. 1).

Example 7 (Example 4 cont.). $\mathcal{P}_1$ and $\mathcal{P}_2$ have no infinite positions because for both programs the chase terminates. Consequently, they are *GSCh*. Consider a program $\mathcal{P}_3$ with the same database $D_3 = D_1$ and a set $\mathcal{P}_3^r$ of rules which is $\mathcal{P}_2^r \cup \{\sigma\}$ such that, $\sigma\colon r(X,Y) \to \exists Z\, r(Z,X)$. $r[1]$ and $r[2]$ are infinite positions because, during the chase of $\mathcal{P}_3$, σ cyclically generates infinite null values in $r[2]$ that also propagate to $r[1]$. The chase of $\mathcal{P}_3$ does not have the *gsch-property* and it is not *GSCh* since the value b replaces the repeated body variable Y that only appears in infinite positions ($r[1]$ and $r[2]$) and b does not propagate all the way down during the chase procedure. ∎

4   Selection Functions and Program Classes

The finite positions in the definition of the *gsch-property* are not computable for a given program which makes it impossible to decide if the program has the property. Here, we define selection functions that determine subsets of the finite positions of a program. We replace finite positions in the definition of the *gsch-property* with the results from selection functions in order to define new stickiness properties and program classes.

A *selection function* S (over a schema $\mathcal{R}$) is a function that takes a program $\mathcal{P}$ and returns a subset of *FinPoss*($\mathcal{P}$). Particular functions are $S^\perp$ and $S^\top$, that given a program $\mathcal{P}$, return the empty set and *FinPoss*($\mathcal{P}$), respectively. The latter may not be computable, and depends on the program's data, which is not the case for the former. Π_F also defines a data-independent selection function, S^{rank}, that returns the finite-rank positions (there are finitely many values in them in the chase of $\mathcal{P}$, for any data set [10, Lemma 5.1]). A selection function is "syntactically computable" if it only depends on the rules $\mathcal{P}^r$ of a program $\mathcal{P}$, and we use the notation $S(\mathcal{P}^r)$.

The *S-stickiness* is defined by replacing the finite positions in the definition of the *gsch-property* with a selection function S: The chase of a program $\mathcal{P}$ has the *S-stickiness property* if the stickiness condition applies only to values replacing the repeated body variables that do not appear in a position of $S(\mathcal{P})$. *SCh*(S) is the semantic class of programs with the *S-stickiness*. In particular, *SCh* = *SCh*($S^\perp$), *GSCh* = *SCh*($S^\top$). Also, *WSCh* = *SCh*(S^{rank}) is the class of programs with *weak-stickiness of the chase*. *SCh*(S) specifies a range of semantic classes of programs starting with *SCh*, ending with *GSCh*, and with *WSCh* in between.

SCh(S) grows monotonically with S: For selection functions S_1 and S_2 over schema $\mathcal{R}$, if $S_1 \subseteq S_2$, then *SCh*(S_1) $\subseteq$ *SCh*(S_2). Here, $S_1 \subseteq S_2$ if and only if for every program $\mathcal{P}$, $S_1(\mathcal{P}) \subseteq S_2(\mathcal{P})$. In general, the more finite positions are (correctly) identified (and the consequently, the less finite positions are treated as infinite), the more general subclass of *GSCh* that is identified or characterized.

Sticky Datalog$^\pm$ uses the marking procedure to restrict the repeated body variables and impose the *sch-property*. Applying this syntactic restriction only on body variables specified by syntactic selection functions results in syntactic classes that extend sticky Datalog$^\pm$. These syntactic classes are subsumed by the semantic classes defined by the same selection functions; each of these syntactic classes only partially represents its corresponding semantic class. Particularly, *SCh* subsumes sticky Datalog$^\pm$ [10]; and *WS* is a syntactic subclass of *WSCh* (cf. (g) and (h) in Fig. 1).

5   Joint-Weakly-Sticky Programs

The definition of the class of JWS programs uses the syntactic selection function $\mathcal{S}^\exists$, which appeals to the *existential dependency graph* of a program [15] (to define *joint-acyclic* programs). We briefly review it here.

Let $\mathcal{P}^r$ be a set of rules that is standardized apart, i.e. no variable appears in more than one rule. For a variable X, let $B(X)$ $(H(X))$ be the set of all positions where X occurs in the body (head) of its rule σ. For a $\exists$-variable Z, the set of target positions of Z, denoted by $T(Z)$, is the smallest set of positions such that (a) $H(Z) \subseteq T(Z)$, and (b) $H(X) \subseteq T(Z)$ for every $\forall$-variable X with $B(X) \subseteq T(Z)$. Roughly speaking, $T(Z)$ is the set of positions where the null values invented by Z may appear in during the chase.

An *existential dependency graph* (EDG) of $\mathcal{P}^r$ is a directed graph with the $\exists$-variables of $\mathcal{P}^r$ as its nodes. There is an edge from Z to Z' if there exists a body variable X in the rule containing Z' such that $B(X) \subseteq T(Z)$. Intuitively, the edge shows that the values invented by Z might appear in the body of the rule of Z' and cause invention of values by Z'. Therefore, a cycle represents the possibility of infinite null values invention by the $\exists$-variables in the cycle.

Example 8. Let $\mathcal{P}^r$ contain the following rules: $u(Y), r(X,Y) \rightarrow \exists Z\ r(Y,Z)$ and $r(X',Y'), r(Y',Z') \rightarrow p(X',Z')$. For the variable Y, $B(Y) = \{u[1], r[2]\}$, $H(Y) = \{r[1]\}$. Moreover, $T(Z) = \{r[2], p[2]\}$. The EDG of $\mathcal{P}^r$ has Z as its node without any edge since $B(X)$ and $B(Y)$ are not subsets of $T(Z)$. $\mathcal{P}^r$ is not WA, because $r[1]$ and $r[2]$ have infinite rank. ∎

For a set of rules $\mathcal{P}^r$, we define the set of *finite-existential positions* of $\mathcal{P}^r$ denoted by $\Pi_F^\exists(\mathcal{P}^r)$ as follows: It is the set of positions that are not in the target set of any $\exists$-variable in a cycle in $EDG(\mathcal{P}^r)$. Intuitively, a position in $\Pi_F^\exists(\mathcal{P}^r)$ is not in the target of any $\exists$-variable that may invent infinite null values.

Proposition 1. For every set of rules $\mathcal{P}^r$, $\Pi_F(\mathcal{P}^r) \subseteq \Pi_F^\exists(\mathcal{P}^r)$. ∎

$\Pi_F^\exists$ defines a computable selection function $\mathcal{S}^\exists$ that returns finite-existential positions of a program (cf. (c) in Fig. 1). $SCh(\mathcal{S}^\exists)$ is a new semantic subclass of $GSCh$ that generalizes $SCh(\mathcal{S}^{rank})$ since $\mathcal{S}^\exists$ provides a finer mechanism for capturing finite positions in comparison with $\mathcal{S}^{rank}$ (cf. (e) and (f) in Fig. 1).

A program $\mathcal{P}$ is *joint-weakly-sticky* (JWS) if for every rule in $\mathcal{P}^r$ and every variable in its body that occurs more than once, the variable is either non-marked or appears in some positions in $\Pi_F^\exists(\mathcal{P}^r)$. The class of JWS programs is a proper subset of $SCh(\mathcal{S}^\exists)$ and extends WS (cf. (i) and (k) in Fig. 1). Specifically, the program in Example 8 is JWS, because every position is finite-existential, but not WS, because Y' is marked and appears in $r[1]$ and $r[2]$ with infinite rank.

6   A Chase-Based Query Answering Algorithm

SChQA is a QA algorithm for programs in the semantic class of $SCh(\mathcal{S})$. It is based on a bottom-up data generation approach and applies a query-driven chase. The algorithm takes as input a computable selection function $\mathcal{S}$, a program $\mathcal{P} \in SCh(\mathcal{S})$, and a CQ $\mathcal{Q}$ over schema $\mathcal{R}$ and returns $ans(\mathcal{Q}, \mathcal{P})$.

Before describing SChQA, we introduce some notations. A *homomorphism* is a structure-preserving mapping, $h \colon \mathcal{C} \cup \mathcal{N} \to \mathcal{C} \cup \mathcal{N}$, between two instances over schema $\mathcal{R}$ that is the identity on constants. An *isomorphism* is a bijective homomorphism.

Definition 1. A rule $\sigma \in \mathcal{P}^r$ and an assignment θ are *applicable* over an instance I of $\mathcal{R}$ if: (a) $I \models (body(\sigma))[\theta]$; and (b) there is an assignment θ' that extends θ, maps the $\exists$-variables of σ into fresh nulls, and $\theta'(head(\sigma))$ is *not isomorphic* to any atom in I. ∎

Note that for an instance I and a set of rules $\mathcal{P}^r$, we can systematically compute the applicable pairs of rule-assignment by first finding $\sigma \in \mathcal{P}^r$ for which $body(\sigma)$ is satisfied by I. That gives an assignment θ for which $(body(\sigma))[\theta] \in I$. Then, we construct θ' as specified in Definition 1 and we iterate over atoms in I and we check if they are isomorphic to $\theta'(head(\sigma))$.

In SChQA, we use the notion of *freezing a null value* that is moving it from $\mathcal{N}$ into $\mathcal{C}$. It may cause new applicable rule-assignment because it changes isomorphic atoms. Considering an instance I, the *resumption* of a step of SChQA is freezing every null in I and continuing the step. Notice that a pair of rule-assignment is applied only once in Step 2. Moreover, if there are more than one applicable pairs, then SChQA chooses the pair that becomes applicable sooner. SChQA is applicable to any Datalog$^+$ program and any selection function, and returns sound answers. However, completeness is guaranteed only when applied to programs in $SCh(\mathcal{S})$ with a computable $\mathcal{S}$.

Algorithm 1. The SChQA algorithm

Inputs: A selection function $\mathcal{S}$, a program $\mathcal{P} \in SCh(\mathcal{S})$, and a CQ $\mathcal{Q}$ over $\mathcal{P}$.
Output: $ans(\mathcal{Q}, \mathcal{P})$.

Step 1: Initialize an instance I with the extensional database D.

Step 2: Choose an applicable rule-assignment σ and θ over I, add $head(\sigma)[\theta']$ into I in which θ' is an extension of θ with mappings for the $\exists$-variables in σ to fresh nulls in $\mathcal{N}$.

Step 3: Freeze the nulls in the new atom in Step 2 that appear in the positions of $\mathcal{S}(\mathcal{P})$.

Step 4: Iteratively apply Steps 2 and 3 until there is no more applicable pair of rule-assignment.

Step 5: Resume Step 2 with I, i.e. freeze nulls in I and continue with Steps 2. Repeat resumption $M_{\mathcal{Q}}$ times where $M_{\mathcal{Q}}$ is the number of variables in $\mathcal{Q}$

Step 6: Return the tuples in $\mathcal{Q}(I)$ that do not have null values (including the frozen nulls).

Example 9. Consider a program $\mathcal{P}$ with $D = \{s(a, b, c),\ v(b), u(c)\}$, and a BCQ $\mathcal{Q} : p(c, Y) \rightarrow ans_{\mathcal{Q}}$, and a set of rules $\mathcal{P}^r$ containing (the hat signs show the marked variables):

$$\sigma_1 : s(\hat{X}, \hat{Y}, \hat{Z}) \rightarrow \exists W\ s(Y, Z, W). \qquad \sigma_2 : u(\hat{X}) \rightarrow \exists Y, Z\ s(X, Y, Z).$$
$$\sigma_3 : s(\hat{X}, Y, Z), v(\hat{X}), s(Y, Z, \hat{W}) \rightarrow p(Y, Z).$$

$\mathcal{P}$ is in *WS* and so $SCh(\mathcal{S}^{rank})$. Specifically in σ_3, X occurs in $v[1]$ which is in $\mathcal{S}^{rank}(\mathcal{P}^r)$ and Y and Z are not marked. The algorithm starts from $I = D$. At Step 2, σ_1 and $\theta_1 = \{X \rightarrow a, Y \rightarrow b, Z \rightarrow c\}$ are applicable; and SChQA adds $s(b, c, \zeta_1)$ into I. σ_2 and $\theta_2 = \{X \rightarrow c\}$ are also applicable and they add $s(c, \zeta_2, \zeta_3)$ into I. Note that Step 3 does not freeze ζ_1, ζ_2, and ζ_3 since they are not in $\mathcal{S}^{rank}(\mathcal{P}^r)$

There is not more applicable rule-assignments and we continue with Step 5. Consider that σ_1 and $\theta_3 = \{X \rightarrow b, Y \rightarrow c, Y \rightarrow \zeta_1\}$ are not applicable since any $\theta_3' = \theta_3 \cup \{W \rightarrow \zeta_4\}$ generates $s(c, \zeta_1, \zeta_4)$ that is isomorphic with $s(c, \zeta_2, \zeta_3)$ already in I. SChQA is resumed once since $\mathcal{Q}$ has one variable. This is done by freezing $\zeta_1, \zeta_2, \zeta_3$ and returning to Step 2. Now, $s(c, \zeta_1, \zeta_4)$ and $s(c, \zeta_2, \zeta_3)$ are not isomorphic anymore and σ_1 and θ_3 are applied which results in $s(c, \zeta_1, \zeta_4)$. As a consequence, σ_3 and $\theta_4 = \{X \rightarrow b, Y \rightarrow c, Z \rightarrow \zeta_1, W \rightarrow \zeta_4\}$ are applicable, which generate $p(c, \zeta_1)$. The instance I in Step 6 is $I = D \cup \{s(b, c, \zeta_1), s(c, \zeta_2, \zeta_3), s(c, \zeta_1, \zeta_4), p(c, \zeta_1), s(\zeta_2, \zeta_3, \zeta_5), s(\zeta_1, \zeta_4, \zeta_6)\}$, and $I \not\models \mathcal{Q}$. ∎

The number of resumptions with SChQA depends on the query. However, for practical purposes, we could run SChQA with N resumptions, to be able to answer queries with up to N variables. If a query has more than N variables, we can incrementally retake the already-computed instance I, adding the required number of resumptions.

Theorem 1. Consider a computable selection function $\mathcal{S}$, a program $\mathcal{P} \in SCh(\mathcal{S})$, and a CQ $\mathcal{Q}$ over schema $\mathcal{R}$. Algorithm SChQA taking $\mathcal{S}$, $\mathcal{P}$, and $\mathcal{Q}$ as inputs, terminates returning $ans(\mathcal{Q}, \mathcal{P})$. ∎

Termination is due to condition (b) in Definition 1, which prevents isomorphic atoms in I. Note that because of Step 3 the null values that appear in the positions of $\mathcal{S}(\mathcal{P})$ are treated as constants while deciding isomorphic atoms. However, condition (b) in Definition 1 prevents some atoms from I that are necessary for answering $\mathcal{Q}$. Adding these atoms depends on the applicability of certain pairs of rule-assignment in which the assignment replaces some repeated variables in the body of the rule with null values. Each resumption makes some of these pairs applicable by freezing nulls. Since $\mathcal{P}$ is $SCh(\mathcal{S})$, there are at most $M_{\mathcal{Q}}$ such rules and so $M_{\mathcal{Q}}$ resumptions are sufficient for answering $\mathcal{Q}$. The running time of SChQA depends on the number of finite values that may appear in the positions of $\mathcal{S}(\mathcal{P})$.

Proposition 2. Algorithm SChQA runs in PTIME in data if the following holds for $\mathcal{S}$: for any program $\mathcal{P}'$, the number of values appearing in $\mathcal{S}(\mathcal{P}')$-positions during the chase is polynomial in the size of the extensional data. ∎

Lemma 1. During the chase of a Datalog$^+$ program $\mathcal{P}$, the number of distinct values in $\mathcal{S}^\exists(\mathcal{P}^r)$-positions is polynomial in the size of the extensional data. ∎

Corollary 1. SChQA runs in PTIME in data with programs in $SCh(\mathcal{S}^\exists)$, in particular for the programs in the *JWS* and *WS* syntactic classes. ∎

7 Magic-Sets and *JWS* Datalog$^\pm$

Magic-sets is a general technique for rewriting logical rules so that they may be implemented bottom-up in a way that avoids the generation of irrelevant facts [7,11]. The advantage of such a rewriting technique is that, by working bottom-up, we can take advantage of the structure of the query and the data values in it, optimizing the data generation process.

In this section, we present a magic-sets rewriting for Datalog$^+$ programs, denoted by MagicD$^+$. It has two changes regarding the technique in [11] in order to: (a) work with $\exists$-variables in the existential rules, and (b) consider the extensional data of the predicates that also have intensional data defined by the rules. For (a), we apply the solution proposed in [1]. However (b) is specifically relevant for Datalog$^+$ programs that allow predicates with both extensional and intentional data, and we address it in MagicD$^+$. MagicD$^+$ is described in detail in Appendix D in [22].

Example 10 (Example 8 cont.). Consider a BCQ $\mathcal{Q} : p(a, Y) \rightarrow ans_\mathcal{Q}$ over a program $\mathcal{P}$ with $D = \{u(a), r(a, b)\}$ and the rules in $\mathcal{P}^r$. MagicD$^+$ has the following steps:

1. Generate the adorned version of the query by annotating its body predicates with strings of bs and fs that correspond to the positions with constants or variables respectively. Then, propagate the adorned predicates to the other program rules. Here, $p^{bf}(a, Y) \rightarrow ans_\mathcal{Q}$ is the adorned query; $r^{bf}(X, Y), r^{bf}(Y, Z) \rightarrow p^{bf}(X, Z)$ and $u(Y), r^{fb}(X, Y) \rightarrow \exists Z \, r^{bf}(Y, Z)$ are the adorned rules. Note that the first rule in $\mathcal{P}^r$ is not adorned by bounding Z in the head (e.g. $r^{fb}(Y, Z)$) since the $\exists$-variables can not be bounded.
2. Add magic predicates to the body of the adorned rule. The magic predicates specify the values for the bounded variables: $mg\_p^{bf}(X), r^{bf}(X, Y), r^{bf}(Y, Z) \rightarrow p^{bf}(X, Z)$ and $mg\_r^{bf}(Y), u(Y), r^{fb}(X, Y) \rightarrow \exists Z \, r^{bf}(Y, Z)$.
3. Generate magic rules that define the magic predicates: $mg\_p^{bf}(X) \rightarrow mg\_r^{bf}(X)$ and $mg\_r^{bf}(X), r^{bf}(X, Y) \rightarrow mg\_r^{bf}(Y)$, and a fact $mg\_p^{bf}(a)$.
4. For the adorned predicates with extensional data (e.g. r), generate new rules to load their extensional data: $mg\_r^{bf}(X), r(X, Y) \rightarrow r^{bf}(X, Y)$ and $mg\_r^{fb}(Y), r(X, Y) \rightarrow r^{fb}(X, Y)$.

The result is a program $\mathcal{P}_m$ with schema $\mathcal{R}_m$, $D_m = D$, the set of rules $\mathcal{P}_m^r$ specified in Steps 2–5, and $\mathcal{Q}_m$ which is the adorned query from Step 1. ∎

MagicD$^+$ differs from the rewriting algorithm of [1] in Step 4. Particularly, in the latter Step 4 is not needed since, unlike the former, it assumes the intentional predicates in $\mathcal{P}$ and the adorned predicates in $\mathcal{P}_m$ do not have extensional data. Therefore, the correctness of MagicD$^+$, i.e. $ans(\mathcal{Q}, \mathcal{P}) = ans(\mathcal{Q}_m, \mathcal{P}_m)$, follows from both the correctness of the rewriting algorithm in [1] and Step 4.

$\mathcal{P}_m^r$ has certain syntactic properties. First, the magic rules do not have $\exists$-variables. Also as mentioned in Step 1, the positions of $\exists$-variables in the head of a rule never become bounded. Additionally we assume that the full information about bounded variables is propagated from the head of an atom to its body. That is when a variable is in a bounded position in the head it appears in the body only in bounded positions.

Applying MagicD$^+$ over a WS program $\mathcal{P}$, $\mathcal{P}_m$ is not necessarily WS or in $SCh(\mathcal{S}^{rank})$ (cf. Example 14 in Appendix E in [22]), which means $SCh(\mathcal{S}^{rank})$ and WS are not closed under MagicD$^+$. This is because MagicD$^+$ introduces new join variables between the magic predicates and the adorned predicates, and these variables might be marked and appear only in the infinite rank positions. That means the joins may break the $\mathcal{S}^{rank}$-stickiness as it happens in Example 14 in Appendix E [22]. Specifically it turned out to be because $\mathcal{S}^{rank}$ decides some finite positions of $\mathcal{P}_m^r$ as infinite rank positions. In fact, the positions of the new join variables are always bounded and are finite. Therefore, MagicD$^+$ does not break $\mathcal{S}$-stickiness if we consider a finer selection function $\mathcal{S}$ that decides the bounded positions as finite. We show in Theorem 2 that the class of $SCh(\mathcal{S}^\exists)$ and its subclass of JWS are closed under MagicD$^+$ since they apply $\mathcal{S}^\exists$ that better specifies finite positions compared to $\mathcal{S}^{rank}$.

Theorem 2. Let $\mathcal{P}$ and $\mathcal{P}_m$ be the input and the result programs of MagicD$^+$ respectively. If $\mathcal{P}$ is JWS, then $\mathcal{P}_m$ is JWS.   ∎

As a result of Theorem 2, we are able to apply MagicD$^+$ in order to optimize SChQA for the class of JWS and its subclasses sticky and WS.

8   Conclusion and Future Research

We introduced semantic and syntactic extensions of sticky and WS Datalog$^\pm$ and we proposed a practical bottom-up QA algorithm for these programs. We applied a magic-set rewriting technique, MagicD$^+$, to optimize the QA algorithm. As the future work, we intend to study the applications of the magic-set rewriting for Datalog$^\pm$ ontologies and in the presence of program constraints, i.e. negative constraints and equality generating dependencies and specifically for the purpose of managing inconsistency for these ontologies. We believe that SChQA and MagicD$^+$ are applicable on real-world scenarios and we plan to implement them and run experiments on real-world data with large data sets.

References

1. Alviano, M., Leone, N., Manna, M., Terracina, G., Veltri, P.: Magic-sets for datalog with existential quantifiers. In: Proceedings of Datalog 2012, vol. 12, pp. 701–718 (2012)

2. Arenas, M., Gottlob, G., Pieris, A.: Expressive languages for querying the semantic web. In: Proceedings of PODS, pp. 14–26 (2014)
3. Artale, A., Calvanese, D., Kontchakov, R., Zakharyaschev, M.: The DL-lite family and relations. J. Artif. Intell. **36**, 1–69 (2009)
4. Baget, J.F., Leclère, M., Mugnier, M.L., Salvat, E.: Extending decidable cases for rules with existential variables. In: Proceedings of IJCAI, pp. 677–682 (2009)
5. Beeri, C., Vardi, M.Y.: The implication problem for data dependencies. In: Even, S., Kariv, O. (eds.) Automata, Languages and Programming. LNCS, vol. 115, pp. 73–85. Springer, Heidelberg (1981)
6. Beeri, C., Vardi, M.Y.: A proof procedure for data dependencies. J. ACM **31**(4), 718–741 (1984)
7. Beeri, C., Ramakrishnan, R.: On the power of magic. In: Proceedings of PODS, pp. 269–284 (1987)
8. Calì, A., Gottlob, G., Lukasiewicz, T.: A general datalog-based framework for tractable query answering over ontologies. J. Web Semant. **14**, 57–83 (2012)
9. Calì, A., Gottlob, G., Pieris, A.: Ontological query answering under expressive entity-relationship schemata. Inf. Syst. **37**(4), 320–335 (2012)
10. Calì, A., Gottlob, G., Pieris, A.: Towards more expressive ontology languages: the query answering problem. Artif. Intell. **193**, 87–128 (2012)
11. Ceri, S., Gottlob, G., Tanca, L.: Logic Programming and Databases. Springer, Heidelberg (1990)
12. Deutsch, A., Nash, A., Remmel, J.: The chase revisited. In: Proceedings of PODS, pp. 149–158 (2008)
13. Fagin, R., Kolaitis, P.G., Miller, R.J., Popa, L.: Data exchange: semantics and query answering. TCS **336**, 89–124 (2005)
14. Johnson, D.S., Klug, A.: Testing containment of conjunctive queries under functional and inclusion dependencies. In: Proceedings of PODS, pp. 164–169 (1984)
15. Krötzsch, M., Rudolph, S.: Extending decidable existential rules by joining acyclicity and guardedness. In: Proceedings of IJCAI, pp. 963–968 (2011)
16. Leone, N., Manna, M., Terracina, G., Veltri, P.: Efficiently computable datalog$^\exists$ programs. In: Proceedings of KR, pp. 13–23 (2012)
17. Maier, D., Mendelzon, A., Sagiv, Y.: Testing implications of data dependencies. In: Proceedings of TODS, pp. 152–152 (1979)
18. Marnette, B.: Generalized schema-mappings: from termination to tractability. In: Proceedings of PODS, pp. 13–22 (2009)
19. Meier, M., Schmidt, M., Lausen, G.: Efficiently computable datalog$^\exists$ programs. In: Proceedings of VLDB Endowment, pp. 970–981 (2009)
20. Milani, M., Bertossi, L.: Ontology-based multidimensional contexts with applications to quality data specification and extraction. In: Bassiliades, N., Gottlob, G., Sadri, F., Paschke, A., Roman, D. (eds.) RuleML 2015. LNCS, vol. 9202, pp. 277–293. Springer, Heidelberg (2015)
21. Milani, M., Calì, A., Bertossi, L.: A hybrid approach to query answering under expressive datalog$^\pm$. Conference Submission (2016)
22. Milani, M., Bertossi, L.: Extending Weakly-sticky Datalog$^\pm$: Query-answering Tractability and Optimizations. https://goo.gl/bJ8MGA
23. Poggi, A., Lembo, D., Calvanese, D., De Giacomo, G., Lenzerini, M., Rosati, R.: Linking data to ontologies. In: Spaccapietra, S. (ed.) Journal on Data Semantics X. LNCS, vol. 4900, pp. 133–173. Springer, Heidelberg (2008)

A Hybrid Approach to Query Answering Under Expressive Datalog$^{\pm}$

Mostafa Milani[1]([✉]), Andrea Calì[2,3], and Leopoldo Bertossi[1]

[1] School of Computer Science, Carleton University, Ottawa, Canada
{mmilani,bertossi}@scs.carleton.ca
[2] Department of Computer Science, Birkbeck, University of London, London, UK
andrea@dcs.bbk.ac.uk
[3] Oxford-Man Institute of Quantitative Finance, University of Oxford, Oxford, UK

Abstract. Datalog$^{\pm}$ is a family of ontology languages that combine good computational properties with high expressive power. Datalog$^{\pm}$ languages are provably able to capture many relevant Semantic Web languages. In this paper we consider the class of weakly-sticky (WS) Datalog$^{\pm}$ programs, which allow for certain useful forms of joins in rule bodies as well as extending the well-known class of weakly-acyclic TGDs. So far, only nondeterministic algorithms were known for answering queries on WS Datalog$^{\pm}$ programs. We present novel deterministic query answering algorithms under WS Datalog$^{\pm}$. In particular, we propose: *(1)* a bottom-up grounding algorithm based on a query-driven chase, and *(2)* a hybrid approach based on transforming a WS program into a so-called sticky one, for which query rewriting techniques are known. We discuss how our algorithms can be optimized and effectively applied for query answering in real-world scenarios.

1 Introduction

The Datalog$^{\pm}$ family of ontology languages [4], which extends Datalog with explicit existential quantification, has been gaining importance in the area of ontology-based data access (OBDA) due to its capability of capturing several conceptual data models and Semantic Web languages as well as offering efficient query answering services in many variants relevant for applications.

The core feature of Datalog$^{\pm}$ languages are the so-called existential rules (a.k.a. *tuple-generating dependencies* or *TGDs*). Such rules allow the inference (entailment) of new atoms from an initial set of ground atoms, typically a database, through the *chase* procedure [6,7,9]. For example, consider the rule $\forall X \forall Y \, r(X,Y) \rightarrow \exists Z \, s(X,Z)$ (in the following we shall omit universal quantifiers, keeping only the existential ones) and a database D constituted by a single atom $r(a,b)$. A chase step will generate (notice that X and Y correspond to a and b respectively in this step) the atom $s(a, \zeta_1)$, where ζ is a *labelled null*, that is, a placeholder for an unknown value; notice that the constant b is lost in this step as it doesn't appear in the new atom. In general the chase may not terminate. Answering a conjunctive query (CQ) q under a database D and a set

© Springer International Publishing Switzerland 2016
M. Ortiz and S. Schlobach (Eds.): RR 2016, LNCS 9898, pp. 144–158, 2016.
DOI: 10.1007/978-3-319-45276-0_11

of TGDs Σ amounts to computing all atoms entailed by $D \cup \Sigma \cup q$ (where q is seen as a TGD) or, equivalently, evaluating q over all atoms entailed by $D \cup \Sigma$.

Consider the following simple example. Let $\Sigma = \{\sigma_1, \sigma_2\}$; σ_1 is the rule $emp(X) \rightarrow \exists Y\ rep(X, Y)$, asserting that each employee reports to someone; σ_2 is the rule $rep(X, Y) \rightarrow mgr(Y)$, asserting that anyone to whom someone else reports is a manager. Now assume $D = \{mgr(ann), emp(joe)\}$. Let us consider the conjunctive query q_1 defined as $q_1(W_1) \leftarrow rep(W_1, W_2)$, where q_1 is now a new predicate (we use the head predicate q_1 for convenience, but any predicate name would suit). This query, which is itself a TGD[1], which we denote with δ, asks for all those who report to someone. Clearly the plain evaluation $q_1(D)$ over D returns no answer ($q_1(D) = \varnothing$), as D says nothing about who reports to whom. However, we are to reason on atoms entailed by $D \cup \Sigma \cup \delta$; therefore, we apply the chase of D w.r.t. $\Sigma \cup \delta$. By applying σ_1 to $emp(joe)$ we obtain $rep(joe, \zeta_1)$, where ζ_1 is a labeled null (in logical terms this means $\exists Y\ rep(joe, Y)$); then by applying σ_2 to $rep(joe, \zeta_1)$ we obtain $mgr(\zeta_1)$; no more atoms can be entailed by the chase. Now, if we evaluate q_1 on all entailed atoms, thus computing all atoms entailed by $D \cup \Sigma \cup \delta$, we get the answer $\langle joe \rangle$; this is because $D \cup \Sigma \cup \delta$ entails that joe reports to someone: $D \cup \Sigma \cup \delta \models q_1(joe)$. Consider now the query q_2 defined as $q_2(W_1) \leftarrow mgr(W_1)$, asking for all those who are managers. In this case, evaluating q_2 over all atoms entailed by $D \cup \Sigma$ returns the answer $\langle ann \rangle$; notice that $D \cup \Sigma$ does not entail the answer $\langle \zeta_1 \rangle$ for q_2; this because ζ_1 is just a placeholder for an unknown value — we know that the manager to whom joe reports exists, but we do not know who he (or she) is.

Conjunctive query answering under general TGDs is undecidable; languages of the Datalog$^\pm$ family impose therefore restrictions on the form of rules so as to guarantee decidability and certain computational problems, most prominently, conjunctive query answering.

Guarded and *weakly-guarded* Datalog$^\pm$ (the latter generalising the former) were the first decidable Datalog$^\pm$ languages, inspired by guarded logic and characterised by the presence of a *guard atom* in each rule that contains all variables of that rule [5]. The *sticky* Datalog$^\pm$ [5] language was introduced to capture a "proper" notion of *join* in rules, that is, the occurrence of variables in two distinct atoms of a rule body in the absence of a guard for that rule. *Weakly-sticky (WS)* Datalog$^\pm$ extends sticky Datalog$^\pm$ by also capturing the well-known class of *weakly-acyclic* rules [7]. As an example, consider the following set of rules Σ, where some body variables are *marked* (by a hat sign, e.g. $\hat{X}$; see [5]; notice that X and $\hat{X}$ are the same variable) as the result of a procedure that identifies occurrences of variables (the marked ones) corresponding, in the chase procedure, to a value that can eventually be lost in some subsequent chase step.

$$\sigma_1 : v(X) \rightarrow \exists Y\ r(X, Y). \qquad \sigma_3 : r(X, \hat{Y}), r(\hat{Y}, Z) \rightarrow p(X, Z).$$
$$\sigma_2 : p(\hat{X}, \hat{Y}) \rightarrow \exists Z\ p(Y, Z). \qquad \sigma_4 : p(\hat{X}, Y), p(Y, Z) \rightarrow t(Y, Z).$$

A set of TGDs is *sticky* if, for each rule, there is no marked variable in that rule that appears more than once in the body of the same rule. Intuitively,

---

[1] Here we adopt the usual notation for conjunctive queries, where the head appears on the left-hand side.

stickiness can be defined by means of the following (semantic) property: during a chase step according to a rule σ, each value corresponding to a variable appearing more than once in σ is not lost in the chase step, and it is also never lost in any subsequent step involving atoms where it appears. Notice that the set of rules Σ above is not sticky, as easily seen.

Weak acyclicity is defined using the notion of *rank* of a position, i.e. a predicate attribute. In a position of finite (resp., infinite) rank, the number of labelled nulls that can appear in the chase procedure is finite (resp., infinite). In the set of TGDs Σ above, $\Pi_F(\Sigma) = \{v[1], r[1], r[2]\}$ contains the positions with *finite rank* and $\Pi_\infty(\Sigma) = \{p[1], p[2], t[1], t[2]\}$ those with *infinite rank*. A Datalog$^\pm$ program is *weakly-acyclic* if all positions have finite rank. The set Σ of TGDs above is not weakly-acyclic.

A set of TGDs is WS if, for each TGD, every marked variable that appears more than once in the body also appears at least once in a finite-rank position. This notion generalizes both stickiness and acyclicity, because the stickiness condition applies to variables that appear *only* in positions with infinite rank. Notice that Σ above is WS. Specifically, in σ_3, the repeated variable Y appears in positions $r[1]$ and $r[2]$, which are in $\Pi_F(\Sigma)$. In σ_4, Y is repeated not marked.

To answer a conjunctive query q under a set Σ of TGDs (or other types of ontological rules) and a database D, two main approaches were proposed in the literature: *grounding* (or *expansion*) and *query rewriting*. In the grounding approach, variables are suitably replaced by constants (or nulls) in the body of a rule, so that the head of the rule yields a (ground) atom that is entailed by the program. The aforementioned chase is in fact a grounding procedure. The grounding allows the computation of all atoms entailed by $D \cup \Sigma$, onto which q can be then evaluated. In the rewriting approach, the query q is rewritten, according to Σ, into another query q_R (possibly in another language different from that of q), so that the correct answers can be obtained by evaluating q_R directly on D.

The rewriting approach is usually considered more efficient than the grounding because in the former only the query is manipulated, according to the rules, while the data is left unchanged; on the contrary, the grounding approach requires the expansion of the given data, whose size is normally much larger than that of the query and of the rules.

CQ answering can be done in polynomial time in data for WS programs. However, so far, no non-trivial deterministic algorithm for CQ answering has been devised for WS Datalog$^\pm$. In this paper we devise algorithms for the efficient implementation of conjunctive query answering under WS Datalog$^\pm$. Our contributions are as follows.

1. We propose a bottom-up technique, based on grounding, which is a variant of the chase procedure, and relies on a terminating chase-like procedure that is *resumed* a number of times that depends on the query to be evaluated. Once the procedure is resumed a sufficient number of times for the query, the same query is evaluated together with the result of the procedure, that is a set of (ground) rules, yielding the correct answers to the query under the given WS set of TGDs.

2. We propose a *hybrid* approach between grounding and rewriting as follows. First, with a novel algorithm, we transform the given set Σ of TGDs into another, all whose positions in Π_F become of rank 0 (that is, positions where only constants can appear in the chase or grounding). Next, certain variables in such positions are *grounded*, that is, they are replaced by selected constants of the given instance. The obtained program is a sticky program, for which a well-known query rewriting technique for CQ answering can be applied. The rewriting (a union of CQs) is finally evaluated on the initial instance.

Both techniques we propose yield algorithms for CQ answering that are of the same data complexity as the lower-bound complexity of CQ answering under weakly-stick programs. Moreover, both our algorithms, unlike the one for WS Datalog$^{\pm}$ in [5], are deterministic. The advantage of the first, pure-grounding technique is that we can pre-compute off-line a ground Datalog program (possibly containing nulls, treated as constants), which serves for answering every query up to a certain number of variables by simply evaluating it on the minimum model of the above Datalog program (in fact, a relational instance). The second, hybrid algorithm relies on a partial grounding, again computed off-line, as well as on an on-the-fly rewriting of every query; the advantages of this approach are that *(a)* the grounding is query-independent and generally much smaller than a complete grounding; *(b)* the final step consists of a mere evaluation of a union of CQs on the given database. Notice that in both our approaches the last step can be performed by executing an SQL query, then offering the possibility of taking advantage of optimizations of RDBMSs. Several optimization strategies are possible for these algorithms (this is ongoing work). We argue that our techniques set the basis for efficient CQ answering under expressive Datalog$^{\pm}$ languages such as WS Datalog$^{\pm}$. A full version of this paper is available at [14].

2   Preliminaries

In this section, we review some basic notions which we use in the paper.

2.1   Basic Definitions

We assume an infinite universe of data constants Γ_C, an infinite set of (labeled) nulls Γ_N, and an infinite set of variables Γ_V (used in rules and queries). We denote by uppercase letters (e.g. X, Y, Z) variables, while $\boldsymbol{X}$ is a sequence of variables $X_1, \ldots, X_k$ with $k \geq 0$. We use the same notation for sets of variables. A relational schema $\mathcal{R}$ is a finite set of relation names (or predicates). A position $p[i]$ identifies the i-th argument of a predicate p.

A homomorphism is a structure-preserving mapping $h : \Gamma_C \cup \Gamma_N \cup \Gamma_V \to \Gamma_C \cup \Gamma_N \cup \Gamma_V$ such that $c \in \Gamma_C$ implies $h(c) = c$. For atoms and conjunctions of atoms, we denote by Π-homomorphism a homomorphism that is the identity on the terms that appear in a set Π of positions.

A *tuple-generating dependency* (or *TGD*, also called *existential rule*) σ on a schema $\mathcal{R}$ is a formula $p_1(\boldsymbol{X}_1), \ldots, p_n(\boldsymbol{X}_n) \to \exists \boldsymbol{Y} \; p(\boldsymbol{X}, \boldsymbol{Y})$ in which

$p, p_1, \ldots, p_n$ are predicates in $\mathcal{R}$ and $\boldsymbol{X} \subseteq \bigcup \boldsymbol{X}_i$ and $\bigcup \boldsymbol{X}_i$ are universal variables that are implicitly quantified. We denote by $head(\sigma)$ and $body(\sigma)$ the head atom $p(\boldsymbol{X}, \boldsymbol{Y})$ and the set of the body atoms $p_1(\boldsymbol{X}_1), \ldots, p_n(\boldsymbol{X}_n)$ of σ, respectively. Variables in $\boldsymbol{Y}$ are called *existential variables*. A rule is *ground* if its terms are in $\Gamma_C \cup \Gamma_N$.

An instance D for $\mathcal{R}$ is a (possibly infinite) set of atoms with predicates in $\mathcal{R}$ and arguments from $\Gamma_C \cup \Gamma_N$. A database D is an instance that contains only atoms with arguments from Γ_C. The *active domain* of a database D denoted by $active(D)$ is the set of constants that appear in D.

A rule σ is satisfied by an instance I, written $I \models \sigma$, if the following holds: whenever there exists a homomorphism h such that $h(body(\sigma)) \subseteq I$, then there exists a homomorphism h' as an extension of h that maps existential variables of σ into terms in $\Gamma_C \cup \Gamma_N$, such that $h'(head(\sigma)) \subseteq I$. An instance I satisfies a set Σ of TGDs, denoted $I \models \Sigma$, if $I \models \sigma$ for each $\sigma \in \Sigma$.

A conjunctive query (CQ) has the form $q(\boldsymbol{X}) \leftarrow p_1(\boldsymbol{X}_1), \ldots, p_n(\boldsymbol{X}_n)$ where $p_1, \ldots, p_n$ are predicate names in $\mathcal{R}$, q is a predicate name not in $\mathcal{R}$, $\boldsymbol{X} \subseteq \bigcup \boldsymbol{X}_i$ and the $\boldsymbol{X}_i$ are sequences of variables or constants. A Boolean CQ (BCQ) over $\mathcal{R}$ is a CQ having head predicate q of arity 0 (i.e., no variables in $\boldsymbol{X}$). The answer to a CQ $q(\boldsymbol{X}) \leftarrow p_1(\boldsymbol{X}_1), \ldots, p_n(\boldsymbol{X}_n)$ over an instance I, denoted as $q(I)$, is the set of all n-tuples $t \in \Gamma_C^n$ for which there exists a homomorphism h such that $h(p_1(\boldsymbol{X}_1), \ldots, p_n(\boldsymbol{X}_n)) \subseteq I$ and $h(\boldsymbol{X}) = t$. A BCQ has only the empty tuple $\langle \rangle$ as possible answer, in which case we say it has a positive answer, denoted $I \models q$.

A program $\mathcal{P}$ consists of a set of rules $\Sigma^{\mathcal{P}}$ and a database $D^{\mathcal{P}}$ over same schema $\mathcal{R}$. $\mathcal{P}$ is a ground program if $\Sigma^{\mathcal{P}}$ is a set of ground rules. Given a program $\mathcal{P}$ and a CQ q, the answers to q are those that are true in all models of $\mathcal{P}$. Formally, the models of $\mathcal{P}$, denoted as $mods(\mathcal{P})$, is the set of all instances I such that $I \supseteq D^{\mathcal{P}}$ and $I \models \Sigma^{\mathcal{P}}$. The answers to a CQ q over $\mathcal{P}$, denoted as $ans(q, \mathcal{P})$, is the set of n-tuples $\{t \mid t \in q(I), \forall I \in mods(\mathcal{P})\}$. The answer to a BCQ q is positive, denoted $\mathcal{P} \models q$, if $ans(q, \mathcal{P})$ is not empty.

2.2   Chase and Grounding

The *chase* procedure is a fundamental algorithm in various database problems including implication of database dependencies, query containment and CQ answering under dependencies [3,9,12]. The chase has been broadly employed in CQ answering in the presence of dependencies [5,7]; the intuition is that, given a set of dependencies over a database schema and a fixed database instance as input, the chase "repairs" the instance so that the result satisfies the constraints. The result of the chase procedure, also called chase, is a so-called *universal model* [7], i.e., a representative of all models in $mods(\mathcal{P})$; therefore, the answers to a CQ q under dependencies (in the open-world assumption, also called *certain answers*), can be computed by evaluating q over the chase (and discarding the answers containing labeled nulls). The chase under TGDs, which we do not describe in detail, is in fact a form of *grounding*; in Sect. 3 we propose a grounding technique for answering CQs under WS Datalog$^{\pm}$ based on a variant of the chase.

2.3   Datalog$^\pm$ and the Stickiness Paradigm

Query answering with respect to a set of TGDs is generally undecidable as proved in [2]. The Datalog$^\pm$ family contains syntactic classes of TGDs that impose restrictions on the form of the rules to guarantee decidability and in many cases tractability of query answering. Two relevant decidability paradigms are *guardedness* and *stickiness*. In this paper we concentrate on stickiness, which is a syntactic condition on the join variables in the body of the rules.

Sticky Programs. Sticky rules are defined by means of a body variable *marking procedure* that takes as input the set $\Sigma^\mathcal{P}$ of rules. It has two steps:

1. *Preliminary step*: for each $\sigma \in \Sigma^\mathcal{P}$ and for each variable $X \in body(\sigma)$, if there is an atom $a \in head(\sigma)$ such that X does not appear in a, mark each occurrence of X in $body(\sigma)$.
2. *Propagation step*: for each $\sigma \in \Sigma^\mathcal{P}$, if a marked variable in $body(\sigma)$ appears at position π, then for every $\sigma' \in \Sigma^\mathcal{P}$ (including σ), mark each occurrence of the variables in $body(\sigma')$ that appear in $head(\sigma')$ in the same position π.

Example 1. Consider a program $\mathcal{P}$, with $\Sigma^\mathcal{P}$ as following set of rules in which the marked variables (denoted by hat signs) after applying the preliminary step:

$$r(X,Y), p(X,Z) \rightarrow s(X,Y,Z). \qquad u(X) \rightarrow \exists Y\, r(Y,X).$$
$$s(\hat{X}, Y, \hat{Z}) \rightarrow u(Y).$$

In the first rule, variables X and Z are marked after applying one propagation step since they appear in the head in marked positions ($s[1], s[3]$), and the final, marked rules are:

$$r(\hat{X}, Y), p(\hat{X}, \hat{Z}) \rightarrow s(X,Y,Z). \qquad u(X) \rightarrow \exists Y\, r(Y,X).$$
$$s(\hat{X}, Y, \hat{Z}) \rightarrow u(Y).$$

∎

$\mathcal{P}$ is sticky when, at the end of the marking procedure over $\Sigma^\mathcal{P}$, there is no rule with a marked variable in its body that occurs more than once. From Example 1, we can see that the program is *not* sticky since X in the first rule is marked and occurs twice in $r[1]$ and $p[1]$.

Sticky programs enjoy *first-order rewritability*. A class of programs is first-order rewritable if, for every program $\mathcal{P}$ in the class, and for every BCQ q, there is a first-order query $q_\mathcal{P}$ such that, $\mathcal{P} \models q$ if and only if $D^\mathcal{P} \models q_\mathcal{P}$. Thus, under first-order rewritable programs, CQs can be answered by constructing the (finite) rewritten first-order query [8], and then evaluating it over the extensional database. Since evaluation of first-order queries is in AC_0 in data complexity [1], it immediately follows that CQ answering under first-order rewritable classes of rules, including sticky ones, is in AC_0.

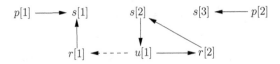

Fig. 1. Dependency graph for $\Sigma^{\mathcal{P}}$ in Example 2

Weakly-Sticky Programs. Weakly-sticky (WS) programs generalize sticky programs and the well known class of weakly-acyclic programs. The class of WS programs is not first-order rewritable but CQ answering over these programs is proved to be tractable in data complexity [5]. The definition of WS programs appeals to conditions on repeated variables in its rule bodies, and is based on the notion of *dependency graph* and the positions with finite rank in such a graph [7] that we explain in Example 2.

Example 2 (Example 1 cont.). Consider the set of rules $\Sigma^{\mathcal{P}}$ in program $\mathcal{P}$. The dependency graph of $\Sigma^{\mathcal{P}}$ is a directed graph constructed as follows: The vertices in V are positions of the predicates in schema of $\Sigma^{\mathcal{P}}$, and the edges in E are defined as it follows. For every $\sigma \in \Sigma^{\mathcal{P}}$ and non-existential variable x in $head(\sigma)$ and in position π in $body(\sigma)$: (1) for each occurrence of x in position π' in $head(\sigma)$, create an edge from π to π'; (2) for each existential variable z in position π'' in $head(\sigma)$, create a *special edge* from π to π''. The dependency graph of $\Sigma^{\mathcal{P}}$ from Example 1 is illustrated in Fig. 1. The special edge from $u[1]$ to $r[1]$ is shown by a dotted arrow that depicts values invention by existential variable Y in the last rule. The *rank* of a position is the maximum number of special edges over all (finite or infinite) paths ending at that position. Accordingly, $\Pi_F(\Sigma^{\mathcal{P}})$ denotes the set of positions of finite rank, and $\Pi_\infty(\Sigma^{\mathcal{P}})$ the set of positions of infinite rank. Intuitively, $\Pi_F(\Sigma^{\mathcal{P}})$ captures positions where finitely many values may appear during the chase; and $\Pi_\infty(\Sigma^{\mathcal{P}})$ those where infinitely many fresh null values may occur during the chase. In this example, $\Pi_\infty(\Sigma^{\mathcal{P}})$ is empty. The rank of $u[1], s[2], r[2], s[3], p[1]$ and $p[2]$ is zero and the rank of $r[1]$ and $s[1]$ is one. ∎

A program $\mathcal{P}$ is WS if for every rule in $\Sigma^{\mathcal{P}}$ and every variable in its body that occurs more that once, the variable is either non-marked or appears at least once in a position in $\Pi_F(\Sigma^{\mathcal{P}})$ (position with finite rank).

Example 3 (Example 2 cont.). $\mathcal{P}$ is WS. $\Pi_F(\Sigma^{\mathcal{P}})$ contains $s[3]$, $p[1]$ and $p[2]$ and the other positions of the predicates in $\Sigma^{\mathcal{P}}$ are in $\Pi_\infty(\Sigma^{\mathcal{P}})$. The repeated variable X in the body of the second rule is marked, but it appears at least once in the finite-rank position $p[1]$. ∎

3   Grounding Based on a Query-Driven Chase

In this section we adopt the chase procedure proposed in [13] for query answering under WS Datalog$^\pm$ programs as a basis for a query-driven grounding algorithm, called GroundWS. GroundWS takes as input a WS Datalog$^\pm$ program $\mathcal{P}$ and a CQ q and returns a ground program $\mathcal{P}'$ for which q can be efficiently answered.

The GroundWS algorithm uses the notion of applicability that we explain next. A rule σ and homomorphism h are *applicable* if: (a) $h(body(\sigma)) \subseteq \mathcal{H} \cup D^{\mathcal{P}}$ in which $\mathcal{H}$ is the set of head atoms of $\Sigma^{\mathcal{P}'}$ (the set of already grounded rules); (b) There is no atom $a \in \mathcal{H}$ and homomorphism h' as an extension of h such that h' maps existential variables in σ to fresh nulls in Γ_N and $h'(head(\sigma))$ is $\Pi_F(\Sigma^{\mathcal{P}})$-homomorphic to a. The second applicability condition is imposed to prevent infinite grounding steps as we describe next. Importantly, we use $\Pi_F(\Sigma^{\mathcal{P}})$-homomorphism instead of ordinary homomorphism to consider only the null values that appear in the positions with infinite rank, which is sufficient to prevent infinite grounding steps. We now illustrate the algorithm GroundWS, which consists of the following steps:

1. Initialize $\Sigma^{\mathcal{P}'}$ to $\varnothing$ and $D^{\mathcal{P}'}$ to $D^{\mathcal{P}}$.
2. For every rule $\sigma \in \Sigma^{\mathcal{P}}$ applicable with homomorphism h, add $h'(\sigma)$ into $\Sigma^{\mathcal{P}'}$, in which h' is extension of h that maps existential variables of σ into fresh null values in Γ_N.
3. Apply Step 2 iteratively, until there are no more applicable rules.
4. Resume Step 2 after freezing every labeled null value in $\Sigma^{\mathcal{P}'}$, where by freezing a null we mean replacing it with a special constant in Γ_C, which henceforth is considered as a constant but is never returned in the result of a query. Repeat this resumption M_q times, where M_q is the number of variables in q.
5. Return $\mathcal{P}'$.

Notice that every pair of rule and homomorphism in Step 2 is applied only once. Moreover, if there are more than one pairs of applicable rule\homomorphism, then GroundWS applies them in a level saturating fashion. More specifically, GroundWS chooses the rule and the homomorphism for which the body atoms have the smallest maximum level. Here the level of an atom is 0 if it is in $D^{\mathcal{P}}$, and it is the maximum level of the body atoms of a rule plus one for the head atom of the rule. It is important to notice that by freezing a null value we consider it as a constant only for deciding homomorphic atoms (specifically in the second applicability condition), and not during query answering. That is, the frozen nulls still can not appear in query answers since they are not in the active domain of the extensional database.

Example 4. Consider WS program $\mathcal{P}$ with $D^{\mathcal{P}} = \{p(a, b), c(b)\}$, and a BCQ $q \leftarrow u(X)$, and a set of rules $\Sigma^{\mathcal{P}}$ as follows:

$$\sigma_1: \quad p(X, Y) \;\rightarrow\; \exists Z \, p(Y, Z). \qquad \sigma_2: \quad p(X, Y), c(X), p(Y, Z) \;\rightarrow\; u(Y).$$

We start from $D^{\mathcal{P}}$ and iteratively generate ground rules by mapping via homomorphism the body of the rules in $\Sigma^{\mathcal{P}}$ into $D^{\mathcal{P}}$ or the head of the rules in $\Sigma^{\mathcal{P}'}$ (the current set of ground rules). The basic algorithm is as follows: we iteratively add a ground rule to $\Sigma^{\mathcal{P}'}$ if its head atom is not homomorphic to the head of a rule already in $\Sigma^{\mathcal{P}'}$, until no new rule can be added. This "cautious" procedure, similar to the chase procedure in [11,13], guarantees termination. In our example, GroundWS stops after adding only one ground rule

$\sigma'_1 : p(a,b) \rightarrow p(b,\zeta_1)$ to $\Sigma^{\mathcal{P}'}$, where ζ_1 is a labelled null in Γ_N. In order to complete our grounding, we *resume* the above basic algorithm M_q times, where M_q is the number of variables in q. Before each resumption, we *freeze* the labelled nulls, which after having frozen are considered as constants. In our example, we have only another resumption, which adds the rules $\sigma''_1 : p(b,\zeta_1) \rightarrow p(\zeta_1,\zeta_2)$ and $\sigma'_2 : p(b,\zeta_1), c(b), p(\zeta_1,\zeta_2) \rightarrow u(\zeta_1)$. Resumptions are needed, intuitively, to capture applications of rules in the chase procedure where a join variable, appearing in two or more distinct atoms, is mapped to a labelled null. ∎

Notice that the number of resumptions depends on the query, which makes our grounding also dependent of the query; however, for practical purposes, we could ground with N resumptions so as to be able to answer queries with up to N existential variables, and if a query with more than N existential variables is to be answered, we can incrementally retake the already-computed grounding and add the required number of resumptions.

Theorem 1. *For every WS program $\mathcal{P}$ and CQ q, GroundWS runs in* PTIME *with respect to the size of $D^{\mathcal{P}}$ and in* 2EXPTIME *with respect to the size of $\mathcal{P}$ and q, and returns a ground program $\mathcal{P}'$ such that, $ans(q,\mathcal{P}) = ans(q,\mathcal{P}')$.*

Proof (sketch): GroundWS runs in polynomial time essentially because of the second applicability condition, which at each resumption prevents the generation in $\Sigma^{\mathcal{P}'}$ of two distinct (ground) rules having $\Pi_F(\Sigma)$-homomorphic heads; since the number of terms in the positions of $\Pi_F(\Sigma)$ is polynomial with respect to $D^{\mathcal{P}}$ [7, Theorem 3.9], the PTIME membership follows. Notice that the above condition, within a resumption "phase", prevents the generation of some (ground) rules in $\Sigma^{\mathcal{P}'}$ that are necessary for answering q; these rules depend on replacing a join variable with a null value in the rule body; the subsequent resumption phase adds at least one of such rules to $\Sigma^{\mathcal{P}'}$. The 2EXPTIME combined complexity is because exponentially many terms can appear in the positions of $\Pi_F(\Sigma)$ with respect to the size of $\mathcal{P}$ and q. This is implicit in the proof of [7, Theorem 3.9]. As a result, the total number of atoms in each resumption phase is double exponential in the size of $\mathcal{P}$ and q.

The weak-stickiness of $\Sigma^{\mathcal{P}}$ implies that such null values continue to appear in the head of consequent ground rules all the way to the head atoms mapped to the query. Hence there are at most M_q such rules generated for answering q, given that each rule "saturates" one of the M_q existential variables in q. Therefore only M_q resumptions are necessary. □

4 Partial Grounding for Weakly-Sticky Datalog$^{\pm}$

In this section we propose a partial grounding algorithm, called PartialGroundingWS, that takes a WS Datalog$^{\pm}$ program $\mathcal{P}$ and transforms it into a sticky Datalog$^{\pm}$ program $\mathcal{P}'$ such that $\mathcal{P}'$ is equivalent to $\mathcal{P}$ for CQ answering. PartialGroundingWS selectively replaces certain variables in positions of finite rank with constants from the active domain of the underlying database. Our algorithm requires that the set of rules $\Sigma^\mathcal{P}$ in the input program satisfies the condition that there is no existential variable in $\Sigma^\mathcal{P}$ in any finite-rank position; therefore each position in the input set of rules $\Sigma^\mathcal{P}$ will have rank either 0 or ∞. The reason for this requirement is the convenience of grounding variables at zero-rank positions by replacing them by constants rather than by labeled nulls. This does not really restrict the input programs since, as we will show later, an arbitrary program can be transformed by the ReduceRank algorithm to a program that has the requirement.

Before illustrating the PartialGroundingWS algorithm, we therefore present an algorithm, called ReduceRank, that takes a program $\mathcal{P}$ and compiles it into an equivalent program $\mathcal{P}'$ whose rule set $\Sigma^{\mathcal{P}'}$ has only zero-rank or infinite-rank positions. The ReduceRank algorithm is inspired by the reduction method in [10] for transforming a weakly-acyclic program into an existential-free Datalog program. Given a program $\mathcal{P}$, ReduceRank executes of the following steps.

1. Initialize $\Sigma^{\mathcal{P}'}$ to $\Sigma^\mathcal{P}$ and $D^{\mathcal{P}'}$ to $D^\mathcal{P}$.
2. Choose a rule σ in $\Sigma^{\mathcal{P}'}$ with an existential variable in a position with rank 1. Notice that if there are existential variables in positions with finite rank, at least one of the positions has rank 1.
3. Skolemize σ as σ' by replacing the existential variable with a functional term. For example, $\sigma : p(X,Y) \rightarrow \exists Z\, r(Y,Z)$, becomes $\sigma' : p(X,Y) \rightarrow r(X,f(X))$.
4. Replace the Skolemized predicate r that has the function term with a new expanded predicate of higher arity (the arity of r plus 1) and introduce a *fresh* special constant of Γ_C to represent the function symbol. The new constant precedes its arguments in a newly introduced position. For example, $r(X,f(X))$ becomes, $r'(X,\mathsf{f},X)$; we are therefore expanding the position $r[2]$.
5. Replace the expanded predicate in other rules and analogously expand other predicates in positions where variables appearing in the expanded position appear: if a variable appears in $\sigma \in \Sigma^\mathcal{P}$ in an expanded position π and also in another position π_1, then also π_1 is expanded (with its predicate) and the same variables of π also appear in π_1, thus preserving the join. More precisely, let X be a variable at position π in an atom $r(\ldots,X,\ldots)$ that is expanded in some rule σ into $r(\ldots,X_1,X_2,\ldots)$; if there is another rule $\sigma_1 \in \Sigma^\mathcal{P}$ containing atoms $r(\ldots,Y,\ldots)$ and $s(\ldots,Y,\ldots)$ such that Y appears in the former at position π_1 and in the latter at position π_2, then the first atom is replaced by $r'(\ldots,Y_1,Y_2,\ldots)$ (expanded version on π_1 with the new variable Y) and the second one is replaced by $s'(\ldots,Y_1,Y_2,\ldots)$ (expanded version on π_2). For example, $r(X,Y),t(Y,Z) \rightarrow s(X,Y,Z)$ becomes $r'(X,Y,Y'),t'(Y,Y',Z) \rightarrow s'(X,Y,Y',Z)$. Notice that if a predicate is expanded in a head-atom in a position where an existential variable occurs, the new positions are not required and are filled with the special symbol
6. If the expanded predicates have extensional data, add new rules to $\Sigma^{\mathcal{P}'}$ to "load" the extensional data into the expanded predicates. For example, if r has extensional data, we add a rule, $r(X,Y) \rightarrow r'(X,Y,\triangle)$. Here, $\triangle$ is used to fill the new position in the expanded predicate since it does not carry extensional data.

7. Repeat Steps 2 to 6 until there is no existential variable in a finite-rank position.

Note that in Step 3 only the body variables that also appear in the head participate as arguments of the function term. For example, in the Skolemization of $p(X, Y) \rightarrow \exists Z \, r(Y, Z)$, the function term does not include X since the rule can be broken down into $p(X, Y) \rightarrow u(Y)$ and $u(Y) \rightarrow \exists Z \, r(Y, Z)$. Notice that given a CQ q over $\mathcal{P}$, Steps 2 to 6 are also applied on q obtaining a new CQ q' over $\mathcal{P}'$.

Example 5. Let $\mathcal{P}$ be a program with $\Sigma^{\mathcal{P}}$ as follows.

$$\sigma_1 : \; v(X) \; \rightarrow \; \exists Y \, r(X, Y). \qquad \sigma_3 : \; t(X, Y), v(X) \; \rightarrow \; p(X, Y).$$
$$\sigma_2 : \; r(X, Y) \; \rightarrow \; \exists Z \, t(X, Z). \qquad \sigma_4 : \; p(X, Y) \; \rightarrow \; \exists Z \, p(Y, Z).$$

In this program, $\Pi_F(\Sigma^{\mathcal{P}}) = \{v[1], r[1], r[2], t[1], t[2]\}$. ReduceRank will eliminate Y in σ_1 and Z in σ_2, but not Z in σ_4 since the later is in an infinite rank position. ReduceRank chooses Y in σ_1 over Z in σ_2 since, Y is in $r[2]$ with zero rank and Z is in $t[2]$ with rank 1. After applying Steps 2–6, σ_1 and σ_2 become $\sigma_1' : \; v(X) \; \rightarrow \; r'(X, \mathsf{f}, X)$ and $\sigma_2' : \; r'(X, Y, Y') \; \rightarrow \; \exists Z \, t(X, Z)$. By removing Y from σ_1, Z in σ_2 is placed in a position with zero rank. ReduceRank repeats Steps 2–6 to eliminate Z in σ_2 which results into $\Sigma^{\mathcal{P}'}$:

$$v(X) \rightarrow r'(X, \mathsf{f}, X). \qquad\qquad r'(X, Y, Y') \rightarrow t'(X, \mathsf{g}, X).$$
$$t'(X, Y, Y'), v(X) \rightarrow p'(X, \triangle, Y, Y'). \quad p'(X, X', Y, Y') \rightarrow \exists Z \, p'(Y, Y', Z, \triangle).$$

Notice that ReduceRank does not try to remove Z in the last rule, since it is in the infinite rank position $p[3]$. Note also that p is expanded twice since both its positions can host labeled nulls generated by Z in σ_2.   ∎

Proposition 1. Given a CQ q over a program $\mathcal{P}$, ReduceRank runs in EXPTIME with respect to the size of $\Sigma^{\mathcal{P}}$ and returns a CQ q' over a program $\mathcal{P}'$ such that $\mathcal{P}'$ has no existential variable in finite rank positions of $\Sigma^{\mathcal{P}'}$ and $ans(q, \mathcal{P}) = ans(q', \mathcal{P}')$.

For every rule in $\Sigma^{\mathcal{P}}$, there is only one corresponding rule in $\Sigma^{\mathcal{P}'}$. There are also rules in $\Sigma^{\mathcal{P}'}$ for loading the extensional data of the expanded predicates. Therefore, the number of rules in $\Sigma^{\mathcal{P}}$ is the same order of the size of $\Sigma^{\mathcal{P}'}$. The arity of the predicates in $\Sigma^{\mathcal{P}'}$ can have an exponential increase with respect to the arity of predicates in $\Sigma^{\mathcal{P}}$ which makes ReduceRank run in EXPTIME. $\mathcal{P}$ and $\mathcal{P}'$ are equivalent since the expanded predicate that represent the propagation of null values in $\mathcal{P}$ are applied on every possible rule in $\Sigma^{\mathcal{P}'}$ in Steps 5–6.

Notice that ReduceRank preserves the weak-stickiness property because the property only concerns repeated marked variables that occur in infinite rank positions, while ReduceRank involves finite rank positions and it does not create a new marked variable or a new infinite rank position to break the property. We can therefore state the following.

Lemma 1. The class of WS programs is closed under ReduceRank.

Now that we explained the ReduceRank algorithm, we continue and present the PartialGroundingWS algorithm. Given a WS program $\mathcal{P}$, let us call *weak rules* the rules of $\Sigma^\mathcal{P}$ in which some repeated marked body variables (which we call weak variables) appear at least once in a position with finite rank. PartialGroundingWS transforms $\mathcal{P}$ into a sticky program $\mathcal{P}'$. The sticky program $\mathcal{P}'$ has the same database as $\mathcal{P}$ ($D^{\mathcal{P}'} = D^\mathcal{P}$) and its set of rules $\Sigma^{\mathcal{P}'}$ is obtained by replacing the weak variables of $\Sigma^\mathcal{P}$ with every constants from the active domain of $D^\mathcal{P}$. Example 6 illustrates the PartialGroundingWS algorithm.

Example 6. Consider a WS program $\mathcal{P}$ with $D^\mathcal{P} = \{p(a,b), r(a,b)\}$ and $\Sigma^\mathcal{P}$ consisting of the following rules:

$$\sigma_1 : p(\hat{X}, \hat{Y}) \;\rightarrow\; \exists Z\, p(Y,Z). \qquad \sigma_3 : s(\hat{X}, Y, Z), r(\hat{X}, Y) \rightarrow t(Y,Z).$$
$$\sigma_2 : p(\hat{X}, Y), p(Y,Z) \;\rightarrow\; s(X,Y,Z).$$

Here σ_3 is a weak rule with X as its weak variable. Notice that Y in σ_2 and σ_3 are not weak since they are not marked (the hat signs show the marked variables). We replace X with constants a and b from $D^\mathcal{P}$. The result is a set of sticky rules $\Sigma^{\mathcal{P}'}$ that includes σ_1 and σ_2 as well as the following rules, $\sigma_3' : s(a,Y,Z), r(a,Y) \rightarrow t(Y,Z)$ and $\sigma_3'' : s(b,Y,Z), r(b,Y) \rightarrow t(Y,Z)$.   ∎

Theorem 2. Let $\mathcal{P}$ be a WS program such that there is no existential variable in $\Sigma^\mathcal{P}$ in a finite rank position. PartialGroundingWS runs in polynomial time with respect to the size of $D^\mathcal{P}$ and it transforms $\mathcal{P}$ into a sticky program $\mathcal{P}'$ such that for every CQ q, the following holds: $ans(q, \mathcal{P}) = ans(q, \mathcal{P}')$.

Proof (sketch): $\mathcal{P}'$ is sticky since every weak variable, that by its definition breaks the weak-stickiness, is grounded. $\mathcal{P}$ and $\mathcal{P}'$ are equivalent for query answering since the weak variables of $\Sigma^\mathcal{P}$ are replaced in $\Sigma^{\mathcal{P}'}$ with every possible constant from $D^\mathcal{P}$, and based on our assumption on $\Sigma^\mathcal{P}$, only constants can substitute these variables. Additionally, PartialGroundingWS runs in polynomial time since weak variables are replaced with constants from $D^\mathcal{P}$.   □

A possible optimization for PartialGroundingWS is to narrow down the values for replacing the weak variables, that is to ignore those constants in the active domain of $D^\mathcal{P}$ that can not appear in the positions where weak variables appear during the chase of $\mathcal{P}$. In Example 6, σ_3' is not useful since a can never be assigned to X in σ_3. For this purpose, GroundWS can be applied to compute the possible values for partial grounding. For example, a CQ $s(X,Y,Z), r(X,Y) \rightarrow q_g(X)$ returns constants for grounding the weak variable X in σ_3.

5   A Hybrid Approach

In this section we propose a query answering algorithm for WS programs based on a hybrid approach that combines ReduceRank and PartialGroundingWS from the previous section with a query rewriting algorithm for sticky programs [8]. Given a WS program $\mathcal{P}$ and a CQ q, hybrid query answering proceeds as follows:

1. Use ReduceRank to compile $\mathcal{P}$ into a WS program $\mathcal{P}'$ with no existential variable in finite rank positions.
2. Apply PartialGroundingWS on $\mathcal{P}'$ that results to a sticky program $\mathcal{P}''$.
3. Rewrite q into a first-order query q' using the rewriting algorithm proposed in [8] and answer q' over $D^{\mathcal{P}''}$ (any other sound and complete rewriting algorithm for sticky programs is also applicable at this step).

Example 7. Consider a WS program $\mathcal{P}$ with database $D = \{v(a)\}$ and $\Sigma^{\mathcal{P}}$ consisting of the following rules:

$$\sigma_1 : p(X,Y) \ \rightarrow \ \exists Z \ p(Y,Z). \qquad \qquad \sigma_4 : r(X,Y), s(X,Z) \rightarrow c(Z).$$
$$\sigma_2 : p(X,Y), p(Y,Z) \ \rightarrow \ u(Y). \qquad \qquad \sigma_5 : c(X) \ \rightarrow \ \exists Y \ p(X,Y).$$

$$\sigma_3 v(X) \rightarrow \ \exists Y \ r(X,Y).$$

The ReduceRank method removes the existential variable Y in σ_3. The result is a WS program $\mathcal{P}'$ with $\Sigma^{\mathcal{P}'}$:

$$p(X,Y) \ \rightarrow \ \exists Z \ p(Y,Z). \qquad r'(X,Y,Y'), s(X,Z) \ \rightarrow \ c(Z).$$
$$p(X,Y), p(Y,Z) \ \rightarrow \ u(Y). \qquad c(X) \ \rightarrow \ \exists Y \ p(X,Y).$$

$$v(X) \ \rightarrow \ r'(X, \mathsf{f}, X).$$

Next, PartialGroundingWS grounds the only weak variable, X in σ_4' with constant a which results into sticky program $\mathcal{P}''$ with $\Sigma^{\mathcal{P}''} = \{\sigma_1, \sigma_2, \sigma_3', \sigma_4'', \sigma_5\}$, in which $\sigma_4'' : r'(a, Y, Y'), s(a, Z) \rightarrow c(Z)$. $\mathcal{P}''$ is sticky and a CQs can be answered by rewriting it in terms of $\Sigma^{\mathcal{P}''}$ and answered directly on $D^{\mathcal{P}''} = D^{\mathcal{P}}$. ∎

Corollary 1. Given a WS program $\mathcal{P}$ and a CQ q, the set of answers obtained from the hybrid approach is $ans(q, \mathcal{P})$.

6 Conclusions

WS Datalog$^\pm$ is an expressive ontology language with good computational properties and capable of capturing the most prominent Semantic Web languages. We proposed two deterministic algorithms for answering conjunctive queries on WS Datalog$^\pm$. In the first algorithm, a variant of the well-known chase, which proceeds in terminating "resumptions", generates an expansion of the given database that contains all (ground) atoms needed to answer the query; the expansion depends on the query as the number of resumptions is the number of existentially quantified variables of the query. For practical purposes, one can expand up to m resumptions off-line, and the expansion will serve to answer all queries with up to m existential variables. If at a certain point a query with more than m existential variables is to be processed, more resumptions can be performed from the expansion already computed. This, of course, if there are no changes in the given database. The second algorithm transforms a WS program into a sticky one by means of (a) a Skolemization and annotation procedure, which turns all finite-rank positions into zero-rank ones, followed by (b) a partial grounding on the zero-ranked positions. Then, the rewriting technique for WS programs is employed; the rewriting, which is in the language of union of conjunctive queries, is then evaluated directly on the given database.

Efficiency. Both algorithms we propose achieve the optimal lower bound in data complexity (i.e., in complexity calculated having only the database as input) for CQ answering under WS Datalog$^{\pm}$, that is PTIME. In the first algorithm, the expansion is computed *off-line*, and the final query processing step is a simple evaluation of a CQ on an instance. In the second algorithm, the rewriting is intensional (i.e., it does involve the data) and the final step is the evaluation of a union of CQs on the given database, which can easily done, for example, by evaluating an SQL query on the database.

In the light of the above considerations, we believe that our contribution sets the basis for practical query answering algorithms in real-world scenarios. We plan to continue our work by running experiments on large data sets. We also intend to refine the hybrid algorithm by limiting the number of CQs in the final rewriting; to do so, we will avoid the grounding of rules when we discover that having certain constants in certain position will not yeld any new atom; such discovery can be performed by analyzing the dependency graph and the TGDs in general. This refinement will improve the efficiency of the algorithm.

References

1. Abiteboul, S., Hull, R., Vianu, V.: Foundations of Databases. Addison-Wesley, Boston (1995)
2. Beeri, C., Vardi, M.Y.: The implication problem for data dependencies. In: Even, S., Kariv, O. (eds.) Automata, Languages and Programming. LNCS, vol. 115, pp. 73–85. Springer, Heidelberg (1981)
3. Beeri, C., Vardi, M.Y.: A procedure for data dependencies. J. ACM **31**(4), 718–741 (1984)
4. Calì, A., Gottlob, G., Lukasiewicz, T.: A general datalog-based framework for tractable query answering over ontologies. J. Web Semant. **14**, 57–83 (2012)
5. Calì, A., Gottlob, G., Pieris, A.: Towards more expressive ontology languages: the query answering problem. Artif. Intell. **193**, 87–128 (2012)
6. Deutsch, A., Nash, A., Remmel, J.B.: The chase revisited. In: Proceedings of PODS, pp. 149–158 (2008)
7. Fagin, R., Kolaitis, P.G., Miller, R.J., Popa, L.: Data exchange: semantics and query answering. TCS **336**, 89–124 (2005)
8. Gottlob, G., Orsi, G., Pieris, A.: Query rewriting and optimization for ontological databases. Proc. TODS **39**(3), 25 (2014)
9. Johnson, D.S., Klug, A.: Testing containment of conjunctive queries under functional and inclusion dependencies. In: Proceedings of PODS, pp. 164–169 (1984)

10. Krötzsch, M., Rudolph, S.: Extending decidable existential rules by joining acyclicity and guardedness. In: Proceedings of IJCAI, pp. 963–968 (2011)
11. Leone, N., Manna, M., Terracina, G., Veltri, P.: Efficiently computable datalog$^{\pm}$ programs. In: Proceedings of KR, pp. 13–23 (2012)
12. Maier, D., Mendelzon, A., Sagiv, Y.: Testing implications of data dependencies. In: Proceedings of TODS, p. 152 (1979)
13. Milani, M., Bertossi, L.: Tractable query answering and optimization for extensions of weakly-sticky datalog+-. In: Proceedings of AMW (2015)
14. Milani, M., Calì, A., Bertossi, L.: A Hybrid Approach to Query Answering under Expressive Datalog$^{\pm}$. Technical report. https://goo.gl/edg9FK

Functional Inferences over Heterogeneous Data

Kwabena Nuamah[(⊠)], Alan Bundy, and Christopher Lucas

School of Informatics, University of Edinburgh, Edinburgh, Scotland, UK
{k.nuamah,c.lucas}@ed.ac.uk, bundy@inf.ed.ac.uk
http://www.inf.ed.ac.uk

Abstract. The increasing availability of knowledge bases (KBs) on the web has opened up the possibility of improved inference in automated query answering (QyA) systems. We have developed a rich inference framework (RIF) that responds to queries where no suitable answer is readily contained in any available data source, by applying functional inferences over heterogeneous data from the web. Our technique combines heuristics, logic and statistical methods to infer novel answers to queries. It also determines what facts are needed for inference, searches for them, and then integrates these diverse facts and their formalisms into a local query-specific inference tree. We explain the internal representation of RIF, the grammar and inference methods for expressing queries and the algorithm for inference. We also show how RIF estimates confidence in its answers, given the various forms of uncertainty faced by the framework.

1 Introduction

Inference enables an agent to create new knowledge from old. Our aim is to apply automatic inference to the semantic web, allowing users to extract new knowledge via queries, and dramatically increase the usefulness of semantic web data sources. RIF does not take natural language text as inputs. We use the acronym *QyA* for query answering, to distinguish it from question answering (QA) systems, which tend to focus on natural language processing (NLP) rather than inference of new facts. We focus on queries that require making predictions based on known facts about the past. We evaluate RIF in the domain of open governance, particularly, demography, education and agriculture. We use data from sources such as Wikidata [1], World Bank Data (WBD) (http://data.worldbank.org) and Geonames (http://www.geonames.org).

Our claim is that *the quality and range of answers generated by a query answering system is significantly improved when we automatically curate data and use rich forms of inference to infer novel knowledge from Semantic Web data and other semi-structured data from the web.* We use the term "rich" to emphasize the fact that the RIF relies on inference methods that go beyond first-order logic. We incorporate higher-order inference, where reasoning about functions expands the range of answers that can be sought. For instance, we can use regression to first construct functions then apply them to make predictions. Answers to most

© Springer International Publishing Switzerland 2016
M. Ortiz and S. Schlobach (Eds.): RR 2016, LNCS 9898, pp. 159–166, 2016.
DOI: 10.1007/978-3-319-45276-0_12

queries in Table 2 can only be inferred by such prediction. Our view on QyA complements NLP-driven approaches by inferring non-trivial answers from readily available facts in KBs by applying such rich forms of inference.

Information retrieval has been explored in different ways using techniques that include NLP and formal logic. Data sources from which answers are sought also range from logically represented facts, to natural language text on the Internet. The Semantic Web (SW) [2] offers practical approaches to representing shared knowledge across multiple domains on the Internet. Public agencies are responding to the initiative for open data by publishing their data using SW technologies. However, most query answering systems are still unable to effectively use these KBs to find answers that require inference beyond the retrieval of facts.

Systems such as AskMSR [4], START [5], Wolfram|Alpha (www.wolframalpha.com), PowerAqua [6], ANGIE [7], and OQA [8] are limited when the required facts are not stored in the KB. GORT [9], although it uses inference, is also heavily dependent on human input of missing facts and does not handle inference over functions. The systems surveyed in the QA tasks of the QALD (Question Answering over Linked Data) [10] challenge do not decompose queries beyond the NLP parse trees. For instance, in [11], the approach taken to answer questions with statistical linked data uses the NLP parse tree to generate the required SPARQL queries. Inference is, therefore, limited to the NLP parse since the process bottoms out at the SPARQL queries that are generated from it. Recent NLP techniques, such as dependency-based compositional semantics (DCS) [12], use statistical techniques that involve semantic parsing of questions to logical forms and evaluation of the logical forms with respect to a database of facts.

We use techniques from SW, logic and statistical inference to build the RIF.

2   The Rich Inference Framework

RIF uses a graph-based algorithm that recursively decomposes queries into sub-queries, eventually grounding out in either stored facts or previously cached answers. The decomposition at each level, as well as the means for combining sub-queries, is determined by features of the query or the sub-query's parent.

Facts retrieved from the external KBs used by the framework are primarily based on RDF [3] and are queried using the SPARQL query language or specific web APIs provided by the sources. This information needs to be curated to enrich it for the inference that is to follow. We augment the *subject(subj)*, *predicate(pred)*, *object(obj)* triple found in RDF KBs with *frames* that contain additional elements such as time, uncertainty, units of quantities, and other features as required. A frame is a list of *key:value* pair elements with keys that include (but are not limited to) *subj, pred, obj, time* and *confidence*. For example, the frame *[method:VALUE, subj:uk, pred:population, obj:63182000, time:2011, confidence:0.35]* represents the population of the UK in 2011 and the confidence RIF has in this fact.

2.1   Definitions

Definition 1. RIF Node: *A RIF node is a frame with elements whose values contain variables or ground terms.*

Variables indicate the elements of the frame to be looked up or inferred. Variables are prefixed with the $ or ? symbols. All variables in a RIF node are bound. Variables whose values are returned from a node are prefixed with the ? symbol. For instance, *[method:MAX, subj:uk, pred:population, obj:?y, time:2020]* shows that the object, y, is unknown and must be inferred and returned. An answer is found when all variables are instantiated to ground terms.

Definition 2. RIF Tree: *A tree of RIF nodes where each child node is derived from a decomposition of its parent node.*

RIF performs inference on a tree in two directions: (1) top-down: decomposing nodes using inference strategies, (2) bottom-up: using inference methods to propagate values from the leaf nodes back up the RIF tree to the root. Decomposition strategies label the arcs. Inference methods label the nodes.

Definition 3. Inference Method: *A higher-order function that aggregates values from a set of RIF nodes.* For instance, for a given node n_{parent}, its child nodes n_{child}^i and inference method, Σ_I, $n_{parent}.obj = \Sigma_I\{n_{child}^i.obj\}$, where $x.obj$ is the *object* element of the RIF node, x.

We use the notation *frame.key* to extract the value of the specified element from the frame of a given RIF *node*.

In RIF, methods applied to RIF also return RIF nodes. An inference method, first extracts relevant values from its child RIF nodes to use as inputs and then applies the function associated with the method. The inference method then substitutes the inferred value into the respective RIF node elements and returns the complete RIF node. Methods used in RIF are listed in Table 1.

Definition 4. Query: *A query is a composition of inference methods and contains both functional and propositional logics for describing entities and relations.*

Functional is used at meta-level for inference methods and propositional for the object logic. We use a context-free grammar in Extended Backus-Naur Form (BNF) that, together with the type signatures of inference methods, defines well-formed queries. RIF queries take the form:

```
func_expr :: METHOD_NAME((var|<var,var>),(logic_expr|func_expr)[,(logic_expr|func_expr)])
```

where *var* are variables, *func_expr* are functional expressions and *logic_expr* represents propositional expressions. Examples are shown in Table 2. The convention is that methods are all caps and propositional constants can begin with either a lower or upper case.

Predicates are not pre-defined prior to their use in RIF. The framework finds matching predicates in the KBs from which the corresponding subjects (or objects) are retrieved. The matching process uses string functions to split words in a predicate, language resources such as WordNet [13] to find synonyms, and edit distance measures to find matches to predicates in a KB.

Table 1. Inference methods

Method	Description
VALUE	Default method. Returns value of node
SUM	Add values of nodes of numeric type
AVG	Mean value of child nodes
MEDIAN	Median value of child nodes
MAX	Maximum value of child nodes
MIN	Minimum value of child nodes
COMP	Obtain a set by list comprehension
GT	'Greater Than' function to compare two nodes
LT	'Less Than' function to compare two nodes
EQ	Check if the value of two nodes are equal
REGRESS	Regression function from child nodes
LOOKUP	Find facts from knowledge bases

Table 2. Types of queries and examples

Type 1: Facts Retrieval

Q1. What was the urban population of UK in 2010?

$VALUE(?y, urban\_population(Uk, ?y, 2010))$

Type 2: Aggregation

Q2. Which country had the lowest female unemployment in South America in 2011?

$MIN(\$y, COMP(\langle ?x, \$y \rangle,$
$\quad female\_unemployment(?x, \$y, 2011):$
$Country(?x)$ & $location(?x, South\_America)))$

Type 3: Nested Queries

Q3. Was the rural population of the country with the largest arable land in Africa greater than the urban population of the country with the smallest arable land in Africa in 2003?

$GT(?b, VALUE(?b, rural\_population(MAX(\$d,$
$COMP(\langle ?c, \$d \rangle, arable\_land(?c, \$d, 2003):$
$Country(?c)$ & $location(?c, Africa))), ?b, 2003)),$
$VALUE(?b, urban\_population(MIN(\$h,$
$COMP(\langle ?g, \$h \rangle, arable\_land(?g, \$h, 2003):Country(?g)$
$\quad$ & $location(?g, Africa))), ?b, 2003)))$

Type 4: Prediction

Q4. What was the GDP in 2010 of the country predicted to have the largest total population in Europe in 2018?

$VALUE(?y, gdp(MAX(\$b, COMP(\langle ?a, \$b \rangle,$
$population(?a, \$b, 2018):Country(?a)$ &
$location(?a, Europe))), ?y, 2010)))$

Definition 5. Inference Strategy: An inference strategy (decomposition) is a transformation on a RIF node from which child nodes are derived.

For a given RIF node, n_{parent} and *strategy*, S, the decomposition, Δ, is the mapping: $\Delta_S(n_{parent}) \mapsto \{n^i_{child} | i > 0\}$.

Strategies used in RIF include: the *temporal* strategy, to decompose nodes by date/time features; *geospatial* strategy, to decompose nodes by location; and the *lookup* strategy, to create child nodes with synonyms of the elements of the parent, to increases the chances of finding facts in KB.

2.2 Implementation

RIF explores strategies and executes the necessary inference methods to infer a novel answer from the available facts. Figure 1 illustrates RIF with different inference strategies that recursively define new nodes and inference methods that

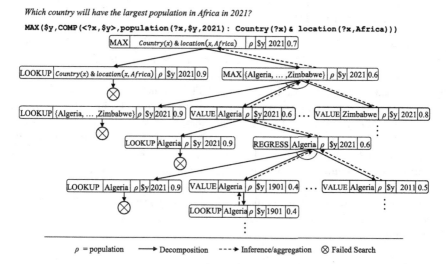

Fig. 1. RIF example: initial decompositions (showing AND-OR search tree). The frame of each node is represented compactly as *[method, subj, pred, obj, time, confidence]*.

infer answers that are propagated up the RIF tree. It is shown as an *AND-OR* graph where the OR branches show strategy options and the AND branches show node decompositions.

RIF begins with the root node as a goal and searches for facts in the KB that match variables in the node. If no fact is found, RIF selects the appropriate strategies to create child nodes. Strategies are selected based on features in queries such as names of location or date/time. For each child node with an answer that is resolved, the answer is propagated back to the parent which in turn aggregates its child nodes and infers a new fact based on its inference method. This process continues until an answer is propagated to the root goal. To prevent unnecessary decompositions, we set a depth limit on the graph. If the inference tree depth bound is reached and no relevant facts are found in the KB at the leaves, the framework yields no answer. For nested queries (e.g. Q3 in Table 2), the parent node spawns child nodes to solve the sub queries.

Due to the time overheads in calling web services, we store a local copy of KBs such as WordNet, ConceptNet [14] and Geonames, which are used frequently by RIF and are rarely changed by their authors. However, data resources such as WBD and the Scottish Government Data (http://statistics.gov.scot/), that are frequently updated, are queried directly using APIs provided by their publishers. In these cases, we cache facts retrieved and inferred as well as inferred functions.

We implemented RIF in Java. We also set up a local RDF triplestore using Apache Jena (https://jena.apache.org) to host frequently used KBs such as Geonames and used a MongoDB (www.mongodb.org) instance for caching.

2.3   Uncertainty in RIF

RIF has two main sources of uncertainty: (1) credibility of KBs and noise in the data, and (2) errors introduced by the inference methods. We have initially focused on the first form of uncertainty and limit it to real-valued facts. We will tackle the latter form in future work. The confidence of a node captures the uncertainty in an estimate, normalized by its magnitude, i.e., the *coefficient of variation (CoV)* or σ/μ, where σ is the estimated standard deviation and μ is the posterior mean. This currently applies to positive real-valued facts. We assume that each retrieved real-valued fact is an observation of the true value with additive Gaussian noise, where the noise variance depends on the assumed credibility of the source. RIF estimates the confidence in an answer by combining and propagating the confidence values of child nodes to their parent recursively in closed form. We use a normal approximation in the estimation and propagation of confidence.

3   Evaluation

We tested our hypothesis by evaluating RIF with a variety of queries (github.com/knuamah/rif). Concretely, we focused on queries that are of interest to public institutions that publish open data on the web. We based the test queries on real-valued facts available in Wikidata and the WBD.

   Existing test sets for evaluating query answering systems focus on aspects of the inference process that differ from our objective with RIF. We were interested in queries that, not only find relevant facts, but also infer non-trivial answers by combining them. We therefore compiled questions that are usually asked about demographics and other country development indicators. Our evaluation consisted of forty queries spanning the four main query types shown in Table 2. Results for the four query types are shown in Table 3. We also used cross-validation to evaluate the confidence scores estimated by RIF. We compared the absolute difference between the inferred value and the true 'held-out' fact to the confidence score (CoV). Results are shown in Fig. 2. We obtained good results in both tests.

   RIF's use of geospatial, temporal and commonsense facts as well as higher order functions allowed it to tackle the range of test queries with 80 % overall success. RIF's main limitation was its word matching mechanism, where it failed

Table 3. Evaluation results by queries types, showing the percentage of queries answered successfully.

Query types	1	2	3	4	Overall
RIF(%)	90	80	80	70	80

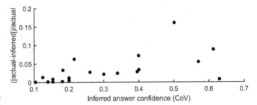

Fig. 2. CoV and estimation error plot.

to find the appropriate matches from KBs in some cases. This was due to its lack of NLP to handle the useful NL descriptions contained in facts. Hence, when the same fact was provided in multiple units, RIF easily mixed them up.

Finally, our evaluation of the confidence scores estimated by RIF also showed a good correlation between the confidence score (CoV) and the error between the true fact and what was inferred.

4   Conclusion

Our Rich Inference approach to QyA enables us to increase the range of queries that can be answered by a QyA system to include prediction and interpolation. The framework also estimates its confidence in the answers inferred given the underlying data and methods used for inference. Finally, the inference trees generated give full access to how answers were inferred. This, we believe, makes our approach practically useful for users who wish to verify answers.

In future work, we plan to extend the algorithm to incorporate the confidence scores in the selection and prioritization of inference strategies, as well as capture non-positive-real-valued data, such as booleans and discrete values, in confidence estimations. We will also consider uncertainty arising from approximations made by inference methods.

References

1. Vrandečić, D., Krötzsch, M.: Wikidata: a free collaborative knowledgebase. Commun. ACM **57**, 78–85 (2014). Springer
2. Berners-Lee, T., Hendler, J.: The semantic web. Sci. Am. **284**, 34–43 (2001)
3. Beckett, D., McBride, B.: RDF/XML syntax specification (revised). W3C recommendation, vol. 10 (2004)
4. Banko, M., Brill, E., Dumais, S., Lin, J.: AskMSR: question answering using the Worldwide Web. In: Proceedings of 2002 AAAI Spring Symposium on Mining Answers, pp. 1–2 (2002)
5. Katz, B.: Annotating the World Wide Web using natural language. In: Proceedings of the 5th RIAO Conference on Computer Assisted Information Searching on the Internet (RIAO 1997) (1997)
6. Lopez, V., Fernández, M., Motta, E., Stieler, N.: PowerAqua: supporting users in querying and exploring the semantic web. Semant. Web **3**, 249–265 (2012)
7. Preda, N., Kasneci, G.: Active knowledge: dynamically enriching RDF knowledge bases by web services. In: SIGMOD, Indianapolis (2010)
8. Fader, A., Zettlemoyer, L., Etzioni, O.: Open question answering over curated and extracted knowledge bases. In: Proceedings of the 20th ACM SIGKDD International Conference on Knowledge Discovery and Data Mining - KDD 2014, pp. 1156–1165. ACM Press (2014)
9. Bundy, A., Sasnauskas, G., Chan, M.: Solving guesstimation problems using the semantic web : four lessons from an application. Semantic Web **6**(2), 197–210 (2015)
10. Unger, C., et al.: Question answering over linked data (QALD-5). In: Working Notes for CLEF 2015 Conference (2015)

11. Höffner, K., Lehmann, J.: Towards question answering on statistical linked data. In: Proceedings of the 10th International Conference on Semantic Systems. ACM (2014)
12. Liang, P., Jordan, M., Klein, D.: Learning dependency-based compositional semantics. Comput. Linguist. **39**, 389–446 (2013). MIT Press
13. Miller, G.A.: WordNet: a lexical database for English. Commun. ACM **38**, 39–41 (1995). ACM
14. Liu, H., Singh, P.: ConceptNet - a practical commonsense reasoning tool-kit. BT Technol. J. **22**, 211–226 (2004). Springer

A Combined Approach to Incremental Reasoning for EL Ontologies

Yuan Ren, Jeff Z. Pan$^{(\boxtimes)}$, Isa Guclu, and Martin Kollingbaum

Department of Computing Science, University of Aberdeen, Aberdeen, UK
jeff.z.pan@abdn.ac.uk

Abstract. Due to the dynamic nature of knowledge and data in semantic applications, ontology incremental reasoning technologies are essential for ontology management systems. Nowadays, many proposed incremental reasoning solutions and implemented systems apply forward chaining completion algorithms to handle the removal and addition of axioms. In this paper, we propose a novel approach to ontology incremental reasoning that combines forward and backward chaining completion for $\mathcal{EL}$. Compared to existing work, this approach can be applied with or without bookkeeping, does not affect parallelisation or tractability, and reduces the effort for re-deriving the over-deleted results both theoretically and empirically.

1 Introduction

Ontologies are widely used in many different application domains to support knowledge management. In order to facilitate automatic processing of ontologies, today's *de facto* standard ontology languages, the Web Ontology Languages (OWLs), are based on a family of Description Logics [1] (DLs). There are some profiles for OWL 2, including EL, QL and RL. Using DLs, ontologies can be regarded as a set of logical axioms.

Ontologies and their corresponding reasoning results are usually considered static. However, with the expanding applications of ontologies, such a paradigm has been challenged [13,16]. In many scenarios, ontologies are subject to rapid changes [12].

The dynamics of ontologies have brought many new research challenges, such as the design of new knowledge representation and query languages [3,5], the development of new reasoning services [10,12] and the development of stream benchmarks [18]. In this paper, we are particularly interested in the development of *incremental reasoning* technologies that update reasoning results affected by the updating of the ontology without naively re-computing all results. In order to reuse previously computed results, many existing approaches [4,8,11,17,21] adopted the Delete and Re-derive (DRed) strategy [7]. With DRed, an incremental reasoner first over-estimates and over-deletes the results affected by the deleted original axioms, unaffected results are preserved. It then re-derives the over-deleted results that can be entailed by the preserved axioms. The authors of [20] pointed out that DRed needs to examine all preserved results during

© Springer International Publishing Switzerland 2016
M. Ortiz and S. Schlobach (Eds.): RR 2016, LNCS 9898, pp. 167–183, 2016.
DOI: 10.1007/978-3-319-45276-0_13

re-derivation and proposed to completely avoid re-derivation by using counting to pinpoint the affected results. Despite of the different mechanisms, all these approaches adopt forward chaining consequence-based algorithms to compute results. For Datalog-based systems, Motik et al. [14] proposed a Backward/Forward (B/F) algorithm for reducing the work done by combining backward and forward chaining to efficiently update the materialization incrementally. In their approach, B/F continuously outperformed the DRed algorithm up to a threshold of 12 % updates in the initial ontology

In this paper, we present a novel DRed approach to ontology incremental reasoning by combining forward chaining and backward chaining of consequence-based algorithms. It has several advantages: (1) It can be applied with either bookkeeping or non-bookkeeping. (2) It helps to reduce the volume of over-deletion and re-derivation in the DRed strategy, as we will show both theoretically and empirically. (3) It can be parallelised and allows the use of multiple computational cores with shared main memory, when applied in parallel reasoners. (4) It works for OWL 2 EL and can be applied to any knowledge representation that supports a consequence-based procedure. When applied with a tractable algorithm, our approach is also tractable. In this paper, we will focus on (1) and (2) with OWL 2 EL [2].

2   Background

In this section, we introduce the most relevant notions of syntax and semantics of DLs. See [1] for a more thorough introduction on DLs.

Briefly, an ontology $\mathcal{O}$ is a set of DL axioms, containing a TBox (schema part of $\mathcal{O}$) and ABox (data part of $\mathcal{O}$). An axiom α is *entailed* by an ontology $\mathcal{O}$, written $\mathcal{O} \models \alpha$, *iff* all models of $\mathcal{O}$ satisfy α. $J_{\mathcal{O}}(\alpha) \subseteq \mathcal{O}$ is a *(minimal) justification* of α iff $J_{\mathcal{O}}(\alpha) \models \alpha$ and $J' \not\models \alpha$ for all $J' \subset J_{\mathcal{O}}(\alpha)$. The algorithms presented in this paper handle both TBox and ABox axioms.

A consequence-based algorithm usually consists of two closely related components: a set of completion rules and a serialised forward-chaining procedure to apply the rules. For example, below is a completion rule for the DL $\mathcal{EL}^+$, in which $\sqsubseteq^*$ is the transitive, reflexive closure of $\sqsubseteq$ in $\mathcal{O}$.

$$\mathbf{R}_{\exists} \frac{E \sqsubseteq \exists R.C, C \sqsubseteq D, R \sqsubseteq^* S}{E \sqsubseteq \exists S.D} : \exists S.D \text{ occurs in } \mathcal{O}$$

We say that some axioms in the ontology can be used as premises (consequences) of a rule when they satisfy the syntactic form specified by the premises (consequences) of the rule. When it is clear from context, we also simply call these axioms premises (consequences) of the rule.

Given an ontology, a consequence-based algorithm repeatedly applies all completion rules in the rule set until no more rule can be applied to compute the *completion closure*:

Definition 1 *(Completion Closure). For a set of axioms S and a completion rule set R, the immediate results of applying R on S, denoted by $R(S)$ or $R^1(S)$,* *is the set of axioms that are either in S, or can be derived as consequence from premises in S by a single rule in R.*

Let $R^{n+1}(S) = R(R^n(S))$ for $n \geq 1$, then the completion closure *of S w.r.t.* R, denoted by $R^*(\mathcal{O})$, *is some* $R^n(S)$ *s.t.* $R^n(S) = R(R^n(S))$.

A rule set R converges *if for any* $\mathcal{O}$, $R^*(\mathcal{O})$ *exists.*

It can be show that the following properties hold:

Lemma 1. *Let* $S_{(i)}$ *be sets of axioms, R a set of completion rules and* $n \geq 1$:

$$S_1 \subseteq S_2 \rightarrow R^n(S_1) \subseteq R^n(S_2) \tag{1}$$

$$R^*(R^n(S)) = R^*(S) \tag{2}$$

$$R^*(S_1 \cup S_2) = R^*(R^n(S_1) \cup S_2) \tag{3}$$

The step of deriving consequences from premises using a rule is an *execution* of the rule. The computation of closure can then be described with the help of a list L in the following algorithm *FCC (Forward Chaining Completion)*.

Forward Chaining Completion:

$FCC(L, S, R)$

INPUT: a list of axioms to be processed L, a set of processed axioms S, a completion rule set R

OUTPUT: a set of processed axioms S

1: **while** $L \neq \emptyset$ **do**
2:     get an element $\alpha \in L$
3:     $L := L \setminus \{\alpha\}$, $S := S \cup \{\alpha\}$
4:     **for** each *rule* in R **do**
5:         **if** α can be used as a premise, and all other premises $\alpha_2, \ldots, \alpha_n$ are in S, and consequence β is not in $S \cup L$ **then**
6:             execute *rule* and add β into L
7: **return** S

For each $\alpha \in L$, *FCC* checks if it can be used to execute a rule with other axioms in S to infer $\beta \notin S \cup L$, which implies that β has not been processed or derived yet. If that is the case, the rule will be executed and β will be added into L. In any case, α will be moved from L to S.

It can be shown that $R^*(L) = FCC(L, \emptyset, R)$:

Lemma 2. *If* $R(S) \subseteq S \cup L$, *then* $R^*(S \cup L) = FCC(L, S, R)$.

Since $R^*(\emptyset) \subseteq L \cup \emptyset$, we have $R^*(L) \subseteq FCC(L, \emptyset, R)$. With the above procedure, consequence-based algorithms can be used to perform ontology reasoning such as classification and materialisation.

In this paper, we focus on incremental reasoning. We consider an ontology sequence $(\mathcal{O}_1, t_1), \ldots, (\mathcal{O}_n, t_n)$, in which $\mathcal{O}_i$ are DL ontologies and $t_1 < \cdots < t_n$ are time points. The change from $\mathcal{O}_i$ to $\mathcal{O}_{i+1}$ is an update of the ontology. In this paper, given an $\mathcal{O}_i$ and its following snapshot $\mathcal{O}_{i+1}$, we address the problem of computing the updated completion closure $FCC(\mathcal{O}_{i+1}, \emptyset, R)$. There are two different approaches to solve this problem. One is *Naive Reasoning*, which recomputes all results completely. The other is *Incremental Reasoning*, which attempts to re-use the results of $FCC(\mathcal{O}_i, \emptyset, R)$ to compute $FCC(\mathcal{O}_{i+1}, \emptyset, R)$, without completely re-computing $FCC(\mathcal{O}_i, \emptyset, R)$. The later is the focus of this paper.

3   Technical Motivation

To deal with incremental reasoning, one key challenge is to handle the deletion of original axioms. The authors of [21] first adopted the Delete and Rederive (DRed) strategy [7,19] from traditional data stream management systems and applied it on ontology incremental reasoning. A DRed approach first *over-deletes* all the potential consequences of the original deletion. Other results will be preserved. It then *re-derives* the over-deleted consequences that can be derived by the preserved results. It finally performs reasoning to deal with the new facts, which can be realised with the same mechanism we just introduced. Such a mechanism has also been adopted by all the existing incremental reasoning approaches. Hence, in this paper we will focus on the optimisation of the over-deletion and re-derivation.

Let $\mathcal{O}$ be an ontology, $R^*(\mathcal{O}) = FCC(\mathcal{O}, \emptyset, R)$ be its completion closure w.r.t. R, $Del \subseteq \mathcal{O}$ be a set of axioms to remove. A DRed approach first identifies a set of *valid over-deletion* $OD \subseteq R^*(\mathcal{O})$ w.r.t. Del:

Definition 2 (Valid Over-deletion). *Let $\mathcal{O}$ be an ontology, $Del \subseteq \mathcal{O}$ a set of axioms to remove, R a completion rule set, an over-deletion OD of $R^*(\mathcal{O})$ w.r.t. Del is valid if:*

1. *$\forall \alpha \in R^*(\mathcal{O})$, if for every $\mathcal{J}_\mathcal{O}(\alpha)$ it is true that $\mathcal{J}_\mathcal{O}(\alpha) \cap Del \neq \emptyset$, then $\alpha \in OD$.*
2. *$\forall \alpha \in R^*(\mathcal{O} \setminus Del)$, there is some $\mathcal{J}_\mathcal{O}(\alpha)$ such that $\mathcal{J}_\mathcal{O}(\alpha) \cap OD = \emptyset$.*

The first condition ensures that OD over-deletes all entailments that can only be inferred from some axioms in Del. The second condition ensures that any entailment of $O \setminus Del$ is also entailed by $R^*(\mathcal{O}) \setminus OD$. A DRed approach then re-derives any axiom $\alpha \in OD$ if there is some $\mathcal{J}_\mathcal{O}(\alpha)$ s.t. $\mathcal{J}_\mathcal{O}(\alpha) \cap Del = \emptyset$. Different DRed or non-DRed incremental reasoning approaches differ primarily on how they identify the over-deleted results and how they perform re-derivation. We introduce them w.r.t. their re-derivation mechanism:

- *Global Re-derivation:* There are a few variants, of the global re-derivation approach, including [4,11,17,21]. These global re-derivation DRed approaches have two major limitations: (1) Some of these approaches require bookkeeping, e.g., a TMS [17], to identify the valid over-deletion. Such bookkeepings will impose performance and resource-consumption overhead; (2) The re-derivation has to go through all remaining axioms $R^*(\mathcal{O}) \setminus OD$ to ensure the completeness of results, even if many of them cannot infer further entailments. A TMS is a loopless directed graph in which nodes denote the axioms in

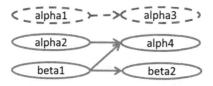

Fig. 1. Over-deletion with a TMS

the completion closure, and edges connect premise and side condition axioms to consequence axioms. When some original axioms are deleted, all axioms to which the deleted axioms have paths in the TMS will be over-deleted. Consider the example:

Example 1. Figure 1 shows a TMS in over-deletion. In this figure, $\mathcal{O} = \{\alpha_1, \alpha_2, \beta_1\}$ and $R^*(\mathcal{O}) = \mathcal{O} \cup \{\alpha_3, \alpha_4, \beta_2\}$.

When $Del = \{\alpha_1\}$ is deleted, $OD = Del \cup \{\alpha_3\}$ since according to the TMS, α_3 is the only entailment connected from α_1 in the TMS.

– *Local Re-derivation:* A recent work [8] proposed a non-bookkeeping DRed approach to address the above limitations. The key point is to exploit the independent nature of different contexts to facilitate the over-deletion and re-derivation. Nevertheless, this approach also has limitations: (1) It relies on the context in rules so it is not applicable to consequence-based algorithms without context; (2) It almost always over-deletes more axioms than necessary. This approach first re-runs a similar forward chaining procedure as in algorithm FCC to identify a set of entailments $DEL = \{C \sqsubseteq D | C \sqsubseteq D \in R^*(\mathcal{O}) \setminus (\mathcal{O} \setminus Del)$ and can be directly or indirectly derived from premises in $Del\}$. The computation of DEL is similar to the over-deletion proposed in [11]. It then computes $Broken = \{C \sqsubseteq E | C \sqsubseteq E \in R^*(\mathcal{O}), C \sqsubseteq D \in DEL\}$, i.e. all derived GCIs who share a LHS (left hand side) context with some GCI that can be derived from the removed axioms. Below is an example:

Example 2. Figure 2 shows a closure similar as the one in Fig. 1. Now the closure is partitioned into two contexts. α_is all have context C_1 and β_is all have context C_2. Suppose we still have $Del = \{\alpha_1\}$ and $DEL = Del \cup \{\alpha_3\}$, since α_2 and α_4 also belongs to the same context, we have $\{\alpha_2, \alpha_4\} \subseteq Broken$.

– *No Re-derivation:* In order to completely avoid the re-derivation in DRed, the authors of [20] apply the counting strategy proposed in [7] to pinpoint the axioms that have to be removed from $R^*(\mathcal{O})$. This approach is conceptually and empirically more efficient than DRed when dealing with removal of axioms.

Example 3. The upper part of Fig. 3 shows a TMS similar as the one in Fig. 1. The main difference is that now the TMS recognised that α_3 can not only be derived from α_1, but also α_2. Hence its count $N(\alpha_3) = 2$.

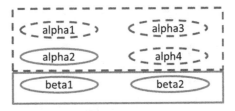

Fig. 2. Over-deletion with context

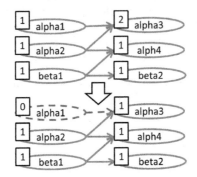

Fig. 3. Over-deletion with counting

The lower part of Fig. 3 shows how the deletion works. When $Del = \{\alpha_1\}$, $N(\alpha_1) = 0$. Consequently $N(\alpha_3) = 1$. Since $N(\alpha_3) \neq 0$, α_3 will not be deleted and its derivation from α_2 is still preserved.

However, it needs to make a trade-off between efficiency and quality of results: (1) If the independent rule executions for each entailment are not thoroughly and precisely identified, then this approach might not yield exactly the same results as naive reasoning. (2) In order to obtain and maintain all possible rule executions for each entailment, this approach essentially computes all justifications for all entailments in the closure. This is known to be expensive even for $\mathcal{EL}$.

In this paper, we do not want to increase the complexity of incremental reasoning in comparison to naive reasoning. We want the results to be exactly the same as naive reasoning. Therefore, we will use DRed instead of the counting strategy. We also want our approach to not to rely on contexts, but be parallelisable with contexts, so that it is applicable with both bookkeeping and non-bookkeeping methods.

4   Combining Forward and Backward Chaining

In order to avoid unnecessary axiom over-deletion and re-derivation, it is necessary to develop a re-derivation mechanism that focuses only on the preserved entailments that can be used to infer over-deleted axioms.

Full Backward Chaining Re-derivation

One way to achieve our goal is a full backward chaining procedure: A backward chaining procedure starts from the entailments that are attempted to be re-derived. It then checks which potential premises in the original closure can be used to derive such an entailment. If all the premises have been preserved during over-deletion or re-derived during re-derivation, then the target axiom can be re-derived. Otherwise, the algorithm can try to re-derive the potential premises recursively. Eventually, this procedure can re-derive all entailments that can be inferred from the preserved axioms. Such a procedure can be described with the following algorithms:

Full Backward Chaining Re-derivation:
$fBCRD(L, S, R)$
INPUT: a list of axioms to be re-derived L, a partial closure S, a completion rule set R
OUTPUT: a set of re-derived axioms $Rederived$

1: $Rederived := \emptyset$
2: **while** $L \neq \emptyset$ **do**
3:     get $\alpha \in L$
4:     $L := L \setminus \{\alpha\}$
5:     $fTest(L, S, Rederived, \{\alpha\}, \alpha, R)$
6: **return** $Rederived$

Given a closure after over-deletion S, a list of over-deleted axioms L and a rule set R, Algorithm $fBCRD(L, S, R)$ finds out all axioms in L that can be directly or indirectly re-derived from S, i.e. $fBCRD(L, S, R) = L \cap R^*(S)$:

1. In Step-1 it first initialises the set of re-derived entailments, which is empty initially.
2. From Step-2 to Step-5, it iteratively tests each entailment $\alpha \in L$ until L is empty. Such a testing is performed by a sub-procedure Algorithm $fTest$.

Full Test:
$fTest(L, S, Rederived, Testing, \beta, R)$
INPUT: a list of axioms to be re-derived L, a partial closure S, a set of re-derived axioms $Rederived$, a set of axioms being tested $Testing$, an axiom to be tested β and a set of completion rules R
OUTPUT: nothing, but L and $Rederived$ will be altered during the execution of the algorithm

1: **if** $\beta \notin Rederived$ **then**
2:     **for** each $rule \in R$ **do**
3:         **if** β can be used as the consequence of $rule$, and all premises $A = \{\alpha_1, \ldots, \alpha_n\}$ of $rule$ are in $S \cup L \cup Rederived \setminus Testing$ **then**

```
4:          while A ∩ L ≠ ∅ do
5:              get α ∈ A ∩ L
6:              fTest(L,S,Rederived,Testing ∪ {β},α,R)
7:          if A ⊆ S ∪ Rederived then
8:              Rederived := Rederived ∪ {β}
9:              L := L \ {β}
10:         return
```

Given S, L, $Rederived$, R and a set of axioms being tested $Testing$, Algorithm $fTest$ will check if an entailment β can be re-derived with R from entailments in S. If β can be re-derived, the algorithm will extend $Rederived$ accordingly. To test the possibility of re-derivation, the algorithm will recursively test the premises of β in L.

Let $R^*(\mathcal{O})$ be a completion closure of ontology $\mathcal{O}$ w.r.t. rule set R and OD be the set of over-deleted axioms, with the above algorithms, re-derivation can be performed with $fBCRD(OD, R^*(\mathcal{O}) \setminus OD, R)$. It can be shown that the above procedure produces the correct and complete re-derivation results:

Lemma 3. *Let $\mathcal{O}$ be an ontology, $S = R^*(\mathcal{O}) = FCC(\mathcal{O}, \emptyset, R)$ be the completion closure of $\mathcal{O}$ w.r.t a set of completion rules R, $Del \subseteq \mathcal{O}$ be a set of deleted axioms, $OD \subseteq S$ be a valid over-deletion w.r.t. Del, then:*

$$(S \setminus OD) \cup fBCRD(OD, S \setminus OD, R) = R^*(\mathcal{O} \setminus Del).$$

Combined Forward and Backward Chaining Re-derivation. The full backward chaining approach introduced in the previous subsection can be further optimised. Particularly, in the presented procedure, the testing of the same axiom may be invoked multiple times.

In this section, we present a more efficient variant of the previous procedure that eliminates the redundant testings. The key-point is to combine forward and backward chaining in re-derivation:

1. Assuming we have a completion closure $S = R^*(\mathcal{O})$ and a set of over-deleted axioms OD, the purpose of re-derivation is to compute $R^*(S \setminus OD)$.
2. Forward chaining re-derivation achieves this by computing $FCC(S \backslash OD, \emptyset, R)$ either globally or locally, and may process unnecessary entailments.
3. Instead, we only need to find $L' = R(S \setminus OD) \setminus (S \setminus OD)$, and then compute $FCC(L', S \setminus OD, R)$. Since we have $R(S \setminus OD) \subseteq (S \setminus OD) \cup L'$, according to Lemma 2, results of $FCC(L', S \backslash OD, R)$ is the same as $R^*((S \backslash OD) \cup L')$. Since $L' = R(S \backslash OD) \setminus (S \backslash OD)$, we have $(S \backslash OD) \cup L' = R(S \backslash OD)$. According to Property (2) of Lemma 1, $R^*((S \backslash OD) \cup L') = R^*(R(S \backslash OD)) = R^*(S \backslash OD)$.

The above procedure is the forward chaining part. The $L' = R(S \setminus OD) \setminus (S \setminus OD)$ will be computed by backward chaining. It can be achieved with a procedure similar to $fBCRD$:

Backward Chaining Re-derivation:

$BCRD(L, S, R)$

INPUT: a list of axioms to be re-derived L, a partial closure S, a completion rule set R

OUTPUT: a set of re-derived axioms $Rederived$

1: $Rederived := \emptyset$
2: **for** each $\alpha \in L$ **do**
3:     $Test(S, Rederived, \alpha, R)$
4: **return** $Rederived$

Test:

$Test(S, Rederived, \beta, R)$

INPUT: a partial closure S, a set of re-derived axioms $Rederived$, an axiom to be tested β and a set of completion rules R

OUTPUT: nothing, but $Rederived$ will be altered during the execution of the algorithm

1: **for** each $rule \in R$ **do**
2:     **if** β can be used as the consequence of $rule$, and all premises $A = \{\alpha_1, \ldots, \alpha_n\}$ of $rule$ are in S **then**
3:         $Rederived := Rederived \cup \{\beta\}$
4:         **return**

As we can see, the procedure is different from the previous full backward chaining re-derivation on the following aspects:

1. Instead of testing axioms with algorithm *fTest*, a new algorithm *Test* is used.
2. *Test* no longer recursively checks if a premise is re-derivable when it is not immediately available in S. Instead, it only checks if all premises are in S, which is the preserved partial closure. Hence the set *Testing* is not needed, because a tested axiom will not be used as premise to re-derive another tested axiom.
3. As a consequence, $BCRD(L, S, R)$ will compute $L \cap R(S)$, namely all axioms in L that can be directly re-derived from S.

Therefore, for a completion closure S and a valid over-deletion OD, we have $R(S \setminus OD) \setminus (S \setminus OD) = BCRD(OD, S \setminus OD, R)$. Combining with the forward chaining part mentioned above, re-derivation of $R^*(S \setminus OD)$ can be achieved:

Theorem 1. *Let $\mathcal{O}$ be an ontology, $S = R^*(\mathcal{O}) = FCC(\mathcal{O}, \emptyset, R)$ be the completion closure of $\mathcal{O}$ w.r.t. a set of completion rules R, $Del \subseteq \mathcal{O}$ be a set of deleted axioms, $OD \subseteq S$ be a valid over-deletion w.r.t. Del, then:*

$$R^*(\mathcal{O} \setminus Del) = FCC(BCRD(OD, S \setminus OD, R), S \setminus OD, R).$$

Proof (Sketch). We can first show that $R^*(\mathcal{O} \setminus Del) = R^*(S \setminus OD)$. Hence we only need to prove $R^*(S \setminus OD) = FCC(BCRD(OD, S \setminus OD, R), S \setminus OD, R)$. According to Lemma 2, we only need to prove that $R(S \setminus OD) \subseteq (S \setminus OD) \cup BCRD(OD, S \setminus OD, R)$. For any $\alpha \in R(S \setminus OD)$, it is either trivially in $(S \setminus OD,$ or it will be added into *Rederived* in Step-2 of Algorithm *Test*. $\qquad\square$

When completion rule set R is tractable, this procedure is also tractable since both $BCRD$ and FCC will be tractable. This suggests that our re-derivation itself does not affect the tractability of reasoning in general. When the completion rule set is intractable, our approach is as complex as forward chaining completion. Although not affecting the worst case computational complexity, conceptually, such a combined forward and backward chaining re-derivation has a minimal problem space (the over-deleted entailments) and a small search space (all premises must be preserved). The re-derivation will only examine the over-deleted axioms *once* in the backward chaining stage and only process the re-derived axioms *once* in the forward chaining stage. These characteristics make the combined re-derivation more efficient than the full forward chaining or full backward chaining re-derivation, especially when the over-deleted entailments are much less than the preserved entailments.

Our approach does not rely on bookkeeping dependencies between premise and consequence axioms. When performing Step-2 of Algorithm *Test*, an implemented system only needs to identify one candidate premise α, and then it can use α in the same way as in Step-5 of Algorithm FCC to find other premises. The identification of α can be realised by exploiting the structural relationships between premise and consequence of each rule. For example, in order to re-derive α (e.g., $E \sqsubseteq \exists S.D$) with backward chaining of rule $\mathbf{R_\exists}$, the reasoner only needs to search for a preserved entailment β (e.g., $E \sqsubseteq \exists R.C$) with the same LHS as α s.t. another γ (e.g., $C \sqsubseteq D$) whose LHS is the RHS filler of β, and whose RHS is the same as α is preserved, and $R \sqsubseteq^* S$ holds. In general, if FCC can be performed without bookkeeping, our approach can be performed without bookkeeping. Nevertheless, our approach can also be augmented with bookkeeping in the same way as FCC. Our approach also does not rely on context in rules. Our approach can also be modified to calculate the counts of entailments, but this is clearly out of the scope of this paper. We will leave it to our future work.

5   Experimental Evaluation

In order to evaluate the usefulness and performance of our approach, we conducted an empirical evaluation to find out:

1. Whether our approach can be used to reduce the number of axioms that are processed, in comparison to global re-derivation and local re-derivation.
2. Whether our approach can be used to achieve efficient incremental reasoning in terms of execution time and memory consumption.

Implementation

For evaluation purposes, we implemented the following approaches:

1. In order to compare with the *naive reasoning* approach, we first implemented a consequence-based algorithm with context. This algorithm is used by the parallel $\mathcal{EL}^{++}$ reasoner ELK [9].
2. In order to compare with the *non-bookkeeping DRed* approach [8], we also implemented an extension of the above approach with the forward-chaining over-deletion used in the non-bookkeeping DRed approach. Such an approach first obtains a set DEL, consisting of all entailments that can be derived from the deleted original axioms. It then effectively over-deletes and re-derives all non-original GCIs OD with the same context as some axiom in DEL.
3. In order to examine the effect of applying our approach with context-based non-bookkeeping DRed, we implemented *our non-bookkeeping approach* on the above one by replacing the context-based re-derivation with our combined forward and backward chaining re-derivation. As we mentioned earlier, this approach will use DEL instead of OD as the over-deleted entailments.
4. In order to compare with the bookkeeping DRed approach, a variant of the first naive reasoning implementation was augmented with the TMS mechanism proposed by [17]. This implementation performs *TMS-based DRed*.
5. In order to examine the effect of applying our approach with TMS-based bookkeeping DRed, we implemented *our TMS-based approach* on the above one by replacing the forward chaining global re-derivation with our re-derivation.

All our implementation used the same completion rules. Hence, they will have the same completion closure for the same input. In order to support reasoning with our evaluation benchmark, our implementations were extended with ABox completion mechanisms. Implementation-wise, this was achieved by internalising ABox axioms with TBox axioms. Such a treatment does not affect the completeness of results on our evaluation benchmark. In order to support the DL used by our evaluation benchmark, our completion rule set extends the **R** rules with the following additional rule to exploit inverse roles in ABox reasoning:

$$\mathbf{R}_{\mathcal{I}} \frac{(a,b):r, a:C}{b:\exists s.C} : \exists s.C \text{ occurs in } \mathcal{O}, r \sqsubseteq^{*,-} s$$

where $r \sqsubseteq^{*,-} s$ if $r \equiv s^- \in \mathcal{O}$ or $r' \sqsubseteq^* r'$, $r \sqsubseteq^{*,-} s'$ and $s' \sqsubseteq^* s$ and $r \sqsubseteq^* s$ if $r \sqsubseteq s \in \mathcal{O}$, or $r \sqsubseteq^* t$ and $t \sqsubseteq^* s$, or $r \sqsubseteq^{*,-} t$ and $t \sqsubseteq^{*,-} s$. With such extension, the rule set is tractable but it is complete for our evaluation benchmark. Note that in the above formulation, all premise axioms still share a context $\{a\}$. Hence, the extension should not affect the context-based parallelisation of the original rule set.

Test Environment

For preparing the evaluation benchmark, we have used the Lehigh University Benchmark (LUBM) [6] with 10 universities, The University Ontology

Benchmark (UOBM)[1] with 10 universities and Systematised Nomenclature of Medicine - Clinical Terms (SNOMED CT)[2] as experimental datasets. We used el-vira[3] to convert UOBM ontologies to OWL 2 EL ontologies.

All experiments were conducted on 64-bit Ubuntu 14.04 with 3.20 GHz CPU and 10G RAM allocated to JVM. To examine if our approach can reduce the number of over-deleted and/or processed axioms in re-derivation, we were interested in the sizes of the following sets:

Del: deleted original axioms.

DEL: the over-deleted non-original axioms directly or indirectly inferred from Del axioms.

R^*: the completion closure.

DEL_L: the non-original axioms directly or indirectly inferred from Del axioms with the forward chaining over-deletion in the non-bookkeeping DRed approach.

OD_L: the non-original axioms with the same context as some axioms in DEL_L. These axioms, even if preserved, will be re-derived by the forward chaining re-derivation of the non-bookkeeping DRed approach.

$BCRD_L$: the axioms re-derived in the backward chaining re-derivation stage of our non-bookkeeping approach.

OD_T: the over-deleted axioms in the TMS-based DRed approach. These are also the axioms to be processed in the backward chaining re-derivation stage of our TMS-based approach.

$BCRD_T$: the axioms re-derived in the backward chaining re-derivation stage of our TMS-based approach. These are also the axioms to be initialised in L in the forward chaining re-derivation stage of our TMS-based approach. Our implementations are available at https://app.box.com/s/mh81cprp0tgpmjc1qmcjdp00powkcpi9.

We conducted the experiments for $n = 1, 2, 5, 10$, i.e. 2%, 4%, 10% and 20% of the ABox were updated respectively. For each n, the size of above sets were obtained on the $\lfloor \frac{150}{n} \rfloor$ runs. The reasoning output of the incremental reasoner was the same as the naive reasoner.

We also explored the performance and memory overhead of the TMS. Naive re-computation was performed by the implementation without TMS. In this experiment, we performed tests for 151 times, on the ABoxes $A_1 \cup \cdots \cup A_{50}, A_2 \cup \cdots \cup A_{51}, \ldots, A_{151} \cup \cdots \cup A_{200}$. Each time, we calculated $\%_{initial}$ and $\%_{memory}$. For $T_{deletion}$ and $T_{addition}$, we conducted the experiments for $n = 1, 2, 5, 10$. For each n, the incremental reasoning were performed for $\lfloor \frac{150}{n} \rfloor$ times. The reasoning output of the incremental reasoner was the same as the naive reasoner.

Test Results

The average percentages of $|Del|$, $|R^* \setminus OD_T|$, $|BCRD_T|$, $|DEL_L|$, $|OD_L|$, $|BCRD_L|$ against $|R^*|$ are illustrated in Table 1.

---

[1] https://www.cs.ox.ac.uk/isg/tools/UOBMGenerator/.

[2] http://www.ihtsdo.org/snomed-ct(2011-Jan.Version).

[3] http://el-vira.googlecode.com.

Table 1. Re-derivation evaluation results (in %)

$\frac{n}{50}$	LUBM				UOBM				SNOMEDCT			
	2	4	10	20	2	4	10	20	2	4	10	20
$|Del|/|R^*|$	0.55	1.10	2.76	5.52	0.72	1.45	3.62	7.26	0.43	0.86	2.16	4.37
$|OD_T|/|R^*|$	1.81	3.60	8.88	17.33	1.61	3.22	8.05	16.13	10.66	20.47	45.72	77.58
$|R^* \setminus OD_T|/|R^*|$	98.2	96.4	91.1	82.7	98.4	96.8	92.0	83.9	89.3	79.5	54.3	22.4
$|BCRD_T|/|R^*|$	0.52	1.02	2.32	4.02	0.03	0.06	0.14	0.28	0.40	0.73	1.60	2.73
$|DEL_L|/|R^*|$	3.66	6.35	13.58	23.86	1.83	3.66	9.15	18.33	5.92	11.75	29.10	58.03
$|OD_L|/|R^*|$	6.47	11.04	22.75	37.88	2.11	4.22	10.57	21.17	5.73	11.42	28.58	58.19
$|BCRD_L|/|R^*|$	1.70	2.59	4.61	6.72	0.03	0.05	0.13	0.27	0.89	1.66	3.78	6.54

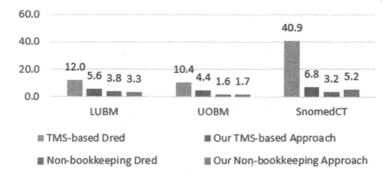

Fig. 4. Time consumption ratio for 2 % Update (in %)

To examine if our approach can be used to achieve efficient incremental reasoning in terms of execution time, in comparison to other approaches, we have conducted experiments using 2 synthetic (LUBM, UOBM) and 1 real-world (SNOMEDCT) datasets to see what would be the ratio of execution time consumed for an update of 2 % in the initial ontology when compared to re-computation. The average values for every approach-dataset pair are illustrated in Fig. 4. We have implemented different algorithms in the environment of TrOWL EL reasoner. Results of experiments are expressed using percentages, instead of absolute values, to proportionally see the effect of different incremental reasoning algorithms and make a comparison between them.

Experiment results regarding the memory overhead are illustrated in Table 2 and Fig. 5.

Observations

1. Because of the nature of *Naive Reasoning*, the cost of time consumed for every small or big update in ontology is always the time of re-computation from scratch(100%). When the update rate is high, this approach can be preferable. But, if the update ratio is as small as 2%, other incremental reasoning techniques become more advantageous. Judging from our experiments, about memory overhead of incremental reasoning, approximately 15% update is the

Table 2. Incremental reasoning evaluation results

$\%_{initial}$	125.89%			
$\%_{memory}$	121.56%			
$\frac{n}{50}$	2%	4%	10%	20%
$\%_{deletion}$	7.37%	14.06%	37.12%	70.21%
$\%_{addition}$	5.94%	15.17%	33.87%	52.44%
$\%_{incremental}$	13.31%	29.23%	70.99%	122.65%

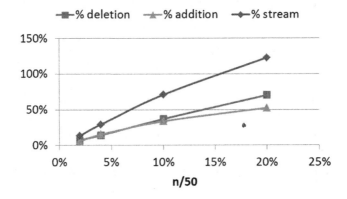

Fig. 5. Incremental reasoning evaluation results

turning point. As illustrated in Table 2 and Fig. 5, up to 15% update in the ontology, incremental reasoning consumes less RAM than naive reasoning, but after that threshold RAM cost of incremental reasoning makes naive reasoning preferable.

2. Using *TMS-based DRed* in a reasoner will impose a performance and memory over-head. The reasoning time was about 25.89 % longer than the same reasoner without TMS. The TMS approach consumed 21.56 % more memory.

3. When *Del* is small, as shown with the ontology SNOMEDCT in Table 1, $BCRD_T$ is much smaller than $R^* \setminus OD_T$ (e.g. 0.40 % v.s. 89.3 % when *Del* is 0.43 % of R^*), indicating that the forward chaining stage in *our TMS-based approach* processes much less axioms than the TMS-based DRed. Even when taking into account the cost of the backward chaining, as implied by the size of OD_T, our combined forward and backward chaining approach should still process less axioms than the TMS-based approach.

4. When the size of DEL_L (non-original axioms directly or indirectly inferred from Del axioms) is smaller than the size of OD_L (axioms that are over-deleted and will be re-derived, even if preserved), the *non-bookkeeping DRed* is unnecessarily over-deleting more entailments than necessary. By applying our non-bookkeeping re-derivation approach, the over-deletion in non-bookkeeping approach can be reduced, i.e. over-deleting DEL_L instead of OD_L.

For example, In case of LUBM with 2% update, 3.66% of data, which constitutes the non-original axioms that are inferred from the deleted original axioms, will be selected for over-deletion. Some of this data will be re-derived in forward chain completion. But non-bookkeeping DRed approach chooses a scope of 6.47% of the data for over deletion. By this way 2.81% of data is unnecessarily processed. In this case our non-bookkeeping approach saves the reasoner from ca.77% (2.81/3.66) of unnecessary processing. In case of UOBM with 2% update, the contribution of our non-bookkeeping DRed approach is 15% ((2.11-1.83)/1.83) when compared to non-bookkeeping DRed approach. In case of SNOMEDCT, we don't observe big contribution but nearly same results.

5. *Our non-bookkeeping approach* and *non-bookkeeping DRed* continuously consumed less computation time when compared to other approaches. When interconnections in ontologies increase, performance advantage of them against naive re-computation and global approach becomes more obvious. Increase in the interconnected axioms makes processing of TMS-based Global DRed longer in terms of execution time but does not have that much increase in the processing of them.

To summarise, our combined forward and backward chaining re-derivation technology is very suitable for ontology updating with small scale deletion. It can significantly reduce the re-derivation effort in comparison to the bookkeeping global re-derivation approach. It can reduce the unnecessary over-deletion in comparison to the non-bookkeeping local re-derivation approach. It can also be used to address the completeness issue of the counting approach.

6 Conclusion

In this paper, we presented a novel approach for ontology incremental reasoning. Although we chose the proposed approach is presented in $\mathcal{EL}$, the approach can be used to other completion based algorithms. The motivation of using $\mathcal{EL}$ is due to the effective $\mathcal{EL}$ based approximate reasoning approach [15] implemented in the TrOWL ontology reasoner. Thus we can combine our approach with the approximate reasoning approach for OWL 2 DL incremental reasoning.

Based on a DRed framework, our approach first uses backward chaining to re-derive the over-deleted axioms that can be directly inferred from preserved axioms, and then uses these directly re-derived axioms to initiate forward chaining and re-derive the completion closure of the preserved axioms. This approach can be combined with different over-deletion techniques. It can also be used with or without bookkeeping. The implementation of our approach does not affect the parallelisation or tractability of reasoning and its mechanism is applicable to many consequence-based algorithm. Evaluation results showed that our approach can indeed reduce unnecessary over-deletion and/or re-derivation in a DRed incremental reasoner and can perform efficient incremental reasoning, particularly when the ontology update is of small size in comparison to the ontology, which is where incremental reasoning is mostly needed.

The backward chaining stage of our approach derives the immediate results of the preserved closure. Such an idea has also been exploited in [11] (in their Algorithm 1.3) and [8] (in their Algorithm 4). The difference is that existing approaches derive such immediate results by forward chaining with all the preserved entailments or un-deleted original axioms, which will essentially re-process the entire new closure or the entire broken contexts. Our approach uses backward chaining to avoid the unnecessary processing. Backward chaining can be implemented easily with rule systems. Hence, the original DRed strategy [7], its declarative variant [19] and the ontological adoption of the latter [21] can also exploit such a backward chaining mechanism. Nevertheless, we notice that backward chaining only needs to be performed to re-derive immediate consequence of the preserved partial closure. Hence, expensive recursive full backward chaining can be avoided. Also, our approach only considers a given completion rule set and does not need to generate additional rules from the axioms.

In the future we would like to combine the strengths of different approaches to develop an adaptive incremental reasoning framework, e.g., using TMS to deal with deletion of side condition axioms and contexts to deal with deletion of non-side condition axioms.

References

1. Baader, F., Calvanese, D., McGuinness, D.L., Nardi, D., Patel-Schneider, P.F. (eds.): The Description Logic Handbook: Theory, Implementation, and Applications. Cambridge University Press, Cambridge (2003)
2. Baader, F., Lutz, C., Suntisrivaraporn, B.: Is tractable reasoning in extensions of the description logic EL useful in practice? In: Proceedings of the 2005 International Workshop on Methods for Modalities (M4M–2005) (2005)
3. Barbieri, D.F., Braga, D., Ceri, S., Valle, E.D., Grossniklaus, M.: C-SPARQL: SPARQL for continuous querying. In: WWW 2009 (2009)
4. Barbieri, D.F., Braga, D., Ceri, S., Della Valle, E., Grossniklaus, M.: Incremental reasoning on streams and rich background knowledge. In: Aroyo, L., Antoniou, G., Hyvönen, E., ten Teije, A., Stuckenschmidt, H., Cabral, L., Tudorache, T. (eds.) ESWC 2010, Part I. LNCS, vol. 6088, pp. 1–15. Springer, Heidelberg (2010)
5. Bolles, A., Grawunder, M., Jacobi, J.: Streaming SPARQL - extending SPARQL to process data streams. In: Bechhofer, S., Hauswirth, M., Hoffmann, J., Koubarakis, M. (eds.) ESWC 2008. LNCS, vol. 5021, pp. 448–462. Springer, Heidelberg (2008)
6. Guo, Y., Pan, Z., Heflin, J.: LUBM: a benchmark for OWL knowledge base systems. Web Semant. Sci. Serv. Agents World Wide Web 3(2–3), 158–182 (2005)
7. Gupta, A., Mumick, I.S., Subrahmanian, V.S.: Maintaining views incrementally. In: SIGMOD 1993 (1993)
8. Kazakov, Y., Klinov, P.: Incremental reasoning in OWL EL without bookkeeping. In: Alani, H., Kagal, L., Fokoue, A., Groth, P., Biemann, C., Parreira, J.X., Aroyo, L., Noy, N., Welty, C., Janowicz, K. (eds.) ISWC 2013, Part I. LNCS, vol. 8218, pp. 232–247. Springer, Heidelberg (2013)
9. Kazakov, Y., Krötzsch, M., Simančík, F.: The incredible ELK. J. Autom. Reasoning 53, 1–61 (2013)

10. Klarman, S., Meyer, T.: Prediction and explanation over DL-*Lite* data streams. In: McMillan, K., Middeldorp, A., Voronkov, A. (eds.) LPAR-19 2013. LNCS, vol. 8312, pp. 536–551. Springer, Heidelberg (2013)
11. Kotowski, J., Bry, F., Brodt, S.: Reasoning as axioms change. In: Rudolph, S., Gutierrez, C. (eds.) RR 2011. LNCS, vol. 6902, pp. 139–154. Springer, Heidelberg (2011)
12. Lecue, F., Pan, J.Z.: Predicting knowledge in an ontology stream. In: Proceedings of the 23rd International Joint Conference on Artificial Intelligence (IJCAI 2013) (2013)
13. Luther, M., Bohm, S., Mobility, S.-A.: An application for stream reasoning. In: Proceedings of 1st International Workshop on Stream Reasoning (SR2009) (2009)
14. Motik, B., Nenov, Y., Piro, R., Horrocks, I.: Incremental update of datalog materialisation: the backward/forward algorithm. In: Proceedings of the 29th National Conference on Artificial Intelligence (AAAI 2015), pp. 1560–1568 (2015)
15. Pan, J.Z., Ren, Y., Zhao, Y.: Tractable approximate deduction for OWL. Artif. Intell. **235**, 95–155 (2016)
16. Parsia, B., Halaschek-Wiener, C., Sirin, E.: Towards incremental reasoning through updates. In: OWL DL, Proceedings of WWW-2006 (2006)
17. Ren, Y., Pan. J.Z.: Optimising ontology stream reasoning with truth maintenance system. In: Proceedings of the 20th ACM International Conference on Information and Knowledge Management, pp. 831–836. ACM (2011)
18. Scharrenbach, T., Urbani, J., Margara, A., Della Valle, E., Bernstein, A.: Seven commandments for benchmarking semantic flow processing systems. In: Cimiano, P., Corcho, O., Presutti, V., Hollink, L., Rudolph, S. (eds.) ESWC 2013. LNCS, vol. 7882, pp. 305–319. Springer, Heidelberg (2013)
19. Staudt, M., Jarke, M.: Incremental maintenance of externally materialized views. In: Vijayaraman, T.M., Buchmann, A.P., Mohan, C., Sarda, N.L. (eds.) Proceedings of the 22th International Conference on Very Large Data Bases (VLDB 1996), 3–6 September 1996, Mumbai, India, pp. 75–86. Morgan Kaufmann (1996)
20. Urbani, J., Margara, A., Jacobs, C., van Harmelen, F., Bal, H.: DynamiTE: parallel materialization of dynamic RDF data. In: Alani, H., Kagal, L., Fokoue, A., Groth, P., Biemann, C., Parreira, J.X., Aroyo, L., Noy, N., Welty, C., Janowicz, K. (eds.) ISWC 2013, Part I. LNCS, vol. 8218, pp. 657–672. Springer, Heidelberg (2013)
21. Volz, R., Staab, S., Motik, B.: Incrementally maintaining materializations of ontologies stored in logic databases. In: Spaccapietra, S., Bertino, E., Jajodia, S., King, R., McLeod, D., Orlowska, M.E., Strous, L. (eds.) Journal on Data Semantics II. LNCS, vol. 3360, pp. 1–34. Springer, Heidelberg (2005)

Short Papers

Society Rules

Abraham Bernstein[✉]

Dynamic and Distributed Information Systems Group, Department of Informatics,
University of Zurich, Zürich, Switzerland
bernstein@ifi.uzh.ch
http://www.ifi.uzh.ch/ddis/people/bernstein.html

Abstract. Our society is full of rules: rules authorize us to achieve our goals by endowing us with legitimation, they provide the necessary structure to understand the chaos of conflicting indications or tell-tales of a situation, and oftentimes they legitimate our actions. But rules in society are different than logical rules suggest to be: they are not as unshakeable, continuously renegotiated, often even accepted to be wrong but still used, and used as inspiration in the situated context rather than universal truth.

Based on theories about the role of technology in society, this talk will first try to convey the role of rules in social science theory. Extending these insights, it will draw on examples to illustrate how they might be transferred to computer science or artificial intelligence to derive systems that are attuned to the role of rules in social environments and adhere to social rules in the environment in which they are used.

Keywords: Rules in the social realm · Non-standard reasoning · Adaptive workflows · Specificity frontier · Process recombination · Cultural adaptivity · Diverse and accurate recommendations

1 Rules in Society

Our Society gets governed by rules. Some are written explicitly such as laws; others are tacit and maintained by processes such as socialization or rites of passage [4]. Many of these rules are used very differently than in the canonical model often-times prescribed by logical rules. They change and evolve during actions [1], are only taken as indications rather than prescriptions for action [10], or are even completely ignored.

Despite this mismatch, the formalization of rules has led to incredible gains: Enterprise Resource Planing Systems (ERPs) such a SAP enable the running of corporations, automated trading systems manage billions, fraud detection systems ensure the stability of our financial transactions. Some of these systems' properties have, however, prevented innovation, caused rigidity, and prevented adaptiveness due to an inability to deal with exceptions or lack of flexibility. In some cases, they may have even led to disasters, as they found themselves in situations that were not foreseen during design and implementation.

© Springer International Publishing Switzerland 2016
M. Ortiz and S. Schlobach (Eds.): RR 2016, LNCS 9898, pp. 187–189, 2016.
DOI: 10.1007/978-3-319-45276-0

2   Social Rules in Systems

Taking inspiration in social science theory about the role of rules and norms in society [6], this talks will explore examples of the loose interpretation of rules as the means for supporting the social rules, norms, or conventions. Each approach presented leverages the use of loosely or statistically specified and interpreted rules in the attempt of finding the sweet-spot between the efficiency of automated interpretation and flexibility of human activity.

The first example will explore an alternative view to process support or workflow management systems that provide flexibility. Based on a concept called the Specificity Frontier [2], it suggests that the relevant rules should be able to change during execution. This has recently lead to a system that interleaves the orchestration of crowds with auto-experimentation to determine the most appropriate process for a given task [3].

The second example will explore the elusive nature of cultural norms—another special set of societal rules—and how they can be leveraged to improve user interactions. Specifically, we show how a rule-based system paired with a very generalizing interpretation of insights from cultural anthropology allow to generate user interfaces that automatically adapt the users' cultural background. These generated user interfaces are shown to increase both the efficiency and effectiveness of users' interactions with the system [7–9].

Time permitting, the third example will take us to the realm of recommending TV shows, where we will see that also statistical reasoning needs to be "bent" to the social rules that govern this specific setting by foregoing recommendation accuracy in favor of diversity and speed [5].

References

1. Barely, S.R.: Technology as an occasion for structuring: evidence from observations of ct scanners and the social order of radiology departments. Adm. Sci. Q. **31**(1), 78–108 (1986)
2. Bernstein, A.: How can cooperative work tools support dynamic group process? bridging the specificity frontier. In: Proceedings of the 2000 ACM Conference on Computer Supported Cooperative Work, CSCW 2000, New York, NY, USA, pp. 279–288. ACM (2000)
3. De Boer, P.M., Bernstein, A.: Pplib: toward the automated generation of crowd computing programs using process recombination and auto-experimentation. ACM Trans. Intell. Syst. Technol. 7(4): 49:1–49:20 (2016)
4. Brown, J.S., Duguid, P.: Organizational learning and communities-ofpractice: toward a unified view of working, learning, and innovation. Organ. Sci. **2**(1), 40–57 (1991)
5. Christoffel, F., Paudel, B., Newell, C., Bernstein, A.: Blockbusters and wallflowers: Accurate, diverse, and scalable recommendations with random walks. In Proceedings of the 9th ACM Conference on Recommender Systems, RecSys 2015, New York, NY, USA, pp. 163–170. ACM (2015)
6. Orlikowski, W.J.: The duality of technology: rethinking the concept of technology in organizations. Organ. Sci. **3**(3), 398–427 (1992)

7. Reinecke, K., Bernstein, A.: Improving performance, perceived usability, and aesthetics with culturally adaptive user interfaces. ACM Trans. Comput.-Hum. Interact. 18(2): 8:1–8:29 (2011)

8. Reinecke, K., Bernstein, A.: Knowing what a user likes: a design science approach to interfaces that automatically adapt to culture. MIS Quarterly **37**(2), 427–453 (2013)

9. Reinecke, K., Nguyen, M.K., Bernstein, A., Näf, M., Gajos, K.Z.: Doodle around the world: online scheduling behavior reflects cultural differences in time perception and group decision-making. In Proceedings of the 2013 Conference on Computer Supported Cooperative Work, CSCW 2013, New York, NY, USA, pp. 45–54. ACM (2013)

10. Suchman, L.A.: Plans and Situated Actions: The Problem of Human-machine Communication. Cambridge University Press, New York (1987)

On the Limits and Possibilities of Query Rewriting

Meghyn Bienvenu[✉]

CNRS, Université de Montpellier and Inria, Montpellier, France
meghyn@lirmm.fr

Recent years have seen an increasing interest in ontology-mediated query answering (OMQA), in which the semantic knowledge provided by an ontology is exploited when querying data. Adding an ontology has several advantages (e.g. simplifying query formulation, integrating data from different sources, providing more complete answers to queries), but it also makes the query answering task more challenging, as reasoning is needed to obtain all answers that can be derived using both the data and the ontology. Query rewriting provides a means of reducing OMQA to the evaluation of database queries (typically, first-order (FO) $\sim$ SQL queries), thereby allowing for OMQA to be built on top of existing database systems and thus to benefit from the maturity and performance of such systems. It is arguably the most prominent algorithmic technique for OMQA. In this talk, I will give an overview of two recent lines of work aimed at understanding the limits and possibilities of query rewriting in OMQA.

The first line of work arose out of the observation that while FO rewritings always exist for ontologies formulated in DL-Lite$_\mathcal{R}$ (the lightweight DL underlying the OWL 2 QL profile), the rewritings generated by implemented rewriting engines were often prohibitively large. This motivated the study of the following succinctness problem: under what circumstances can polynomial-size rewritings be achieved? More specifically, how does the worst-case size of rewritings depend on (i) the way the rewritten queries are represented (e.g. as positive existential queries vs. non-recursive datalog (NDL) queries), (ii) the existential depth of the ontology, and (iii) the structure of the input query (treewidth, number of leaves)? This question has been addressed in a series of works [3, 7, 8], which establish and exploit tight connections between FO query rewriting and circuit complexity. The resulting succinctness landscape shows that while polynomial-size rewritings cannot be guaranteed in general, there are a large classes of ontologies and queries which possess polynomial-size NDL-rewritings. Moreover, concrete NDL-rewriting algorithms that achieve optimal worst-case complexity have recently been developed [4].

At first sight, the FO query rewriting approach seems to have limited applicability, since for almost every ontology language outside the DL-Lite family, we run into the problem that FO rewritings need not exist. However, such results reflect the worst-case situation and leave open the possibility that some, perhaps many, queries encountered in real applications are in fact first-order rewritable. In the second half of this talk, I will give an overview of a recent line of work [1, 2, 5, 6] aimed at devising methods for identifying those ontology-query pairs

© Springer International Publishing Switzerland 2016
M. Ortiz and S. Schlobach (Eds.): RR 2016, LNCS 9898, pp. 190–191, 2016.
DOI: 10.1007/978-3-319-45276-0

which admit FO rewritings, which is an important step towards extending the applicability of the first-order query rewriting approach.

References

1. Bienvenu, M., ten Cate, B., Lutz, C., Wolter, F.: Ontology-based data access: a study through Disjunctive Datalog, CSP, and MMSNP. ACM Trans. Database Syst. (TODS) 39 (2014)
2. Bienvenu, M., Hansen, P., Lutz, C., Wolter, F.: First order-rewritability of conjunctive queries in Horn description logics. In: Proceedings of IJCAI (2016)
3. Bienvenu, M., Kikot, S., Podolskii, V.V.: Tree-like queries in OWL 2 QL: succinctness and complexity results. In: Proceedings of LICS (2015)
4. Bienvenu, M., Kontchakov, R., Kikot, S., Podolskii, V., Zakharyaschev, M.: Theoretically optimal datalog rewritings for OWL 2 QL ontology-mediated queries. In: Proceedings of DL (2016)
5. Bienvenu, M., Lutz, C., Wolter, F.: First order-rewritability of atomic queries in Horn description logics. In: Proceedings of IJCAI, pp. 754–760 (2013)
6. Hansen, P., Lutz, C., Seylan, I., Wolter, F.: Efficient query rewriting in the description logic EL and beyond. In: Proceedings of IJCAI (2015)
7. Kikot, S., Kontchakov, R., Podolskii, V., Zakharyaschev, M.: Exponential Lower Bounds and Separation for Query Rewriting. In: Czumaj, A., Mehlhorn, K., Pitts, A., Wattenhofer, R. (eds.) ICALP 2012, Part II. LNCS, vol. 7392, pp. 263–274. Springer, Heidelberg (2012)
8. Kikot, S., Kontchakov, R., Podolskii, V., Zakharyaschev, M.: On the succinctness of query rewriting over shallow ontologies. In: Proceedings of LICS (2014)

Logic ∧ Reasoning ∧ Scalability ⊨ ⊥?

Ian Horrocks

Department of Computer Science, University of Oxford, Oxford, UK

Logic based "Semantic Technologies" are maturing rapidly, with RDF and OWL now being deployed in diverse application domains, and with major technology vendors starting to augment their existing systems accordingly. For example, the Optique project has successfully piloted Ontology Based Data Access in the energy domain, and Oracle Inc. has enhanced its well-known database management system with modules that use RDF/OWL ontologies to support "semantic data management". Such applications increasingly focus on data, and critically depend on efficient query answering services; this in turn depends on the provision of robustly scalable reasoning systems. In this talk I will review the evolution of Semantic Technologies to date, and show how research ideas from logic based knowledge representation developed into a mainstream technology. I will then go on to examine the scalability challenges arising from deployment in large scale applications, particularly those that primarily focus on query answering over large datasets, compare various different approaches and present some results from ongoing research in the area.

© Springer International Publishing Switzerland 2016
M. Ortiz and S. Schlobach (Eds.): RR 2016, LNCS 9898, p. 192, 2016.
DOI: 10.1007/978-3-319-45276-0

Efficient Computation of Certain Answers: Breaking the CQ Barrier

Leonid Libkin

School of Informatics, University of Edinburgh, Edinburgh, Scotland

Abstract of invited talk: Computing certain answers is the standard way of answering queries over incomplete data; it is also used in many applications such as data integration, data exchange, consistent query answering, ontology-based data access, etc. Unfortunately certain answers are often computationally expensive, and in most applications their complexity is intolerable if one goes beyond the class of conjunctive queries (CQs), or a slight extension thereof.

However, high computational complexity does not yet mean one cannot approximate certain answers efficiently. In this talk we survey several recent results on finding such efficient and correct approximations, going significantly beyond CQs. We do so in a setting of databases with missing values, and first-order (relational calculus/algebra) queries. Even the class of queries where the standard database evaluation produces correct answers is larger than previously thought. When it comes to approximations, we present two schemes with good theoretical complexity. One of them also performs very well in practice, and restores correctness of SQL query evaluation on databases with nulls.

This talk is based on recent papers [1–3].

References

1. Libkin, L.: Certain answers as objects and knowledge. Artif. Intell. **232**, 1–19 (2016)
2. Libkin, L.: SQL's three-valued logic and certain answers. ACM Trans. Database Syst. **41**(1), 1 (2016)
3. Guagliardo, P., Libkin, L.: Making SQL queries correct on incomplete databases: a feasibility study. In: PODS 2016, pp. 211–223

© Springer International Publishing Switzerland 2016
M. Ortiz and S. Schlobach (Eds.): RR 2016, LNCS 9898, p. 193, 2016.
DOI: 10.1007/978-3-319-45276-0

Author Index

Printed in the United States
By Bookmasters